"十二五"职业教育国家规划教材
经全国职业教育教材审定委员会审定

新职业英语 | 行业篇

ENGLISH FOR CAREERS 第三版

机电英语

总主编：徐小贞
主　编：胡庭山
编　者：金　曙　吴　敏　蔡　嵘
　　　　杨茂霞　徐晓燕

外语教学与研究出版社
FOREIGN LANGUAGE TEACHING AND RESEARCH PRESS
北京 BEIJING

图书在版编目（CIP）数据

机电英语／胡庭山主编；金曙等编．— 3 版． — 北京：外语教学与研究出版社，2020.12（2022.12 重印）
（新职业英语行业篇／徐小贞总主编）
ISBN 978-7-5213-2284-2

Ⅰ. ①机… Ⅱ. ①胡… ②金… Ⅲ. ①机电工程－英语－高等职业教育－教材 Ⅳ. ①TH

中国版本图书馆 CIP 数据核字 (2020) 第 261027 号

出 版 人　王　芳
项目负责　姚　瑶
责任编辑　邓　芳
责任校对　武春华
封面设计　孙莉明
版式设计　涂　俐
出版发行　外语教学与研究出版社
社　　址　北京市西三环北路 19 号（100089）
网　　址　http://www.fltrp.com
印　　刷　三河市北燕印装有限公司
开　　本　850×1168　1/16
印　　张　13.5
版　　次　2021 年 1 月第 3 版　2022 年 12 月第 4 次印刷
书　　号　ISBN 978-7-5213-2284-2
定　　价　43.90 元

职业教育出版分社：
地　　址：北京市西三环北路 19 号 外研社大厦 职业教育出版分社 (100089)
咨询电话：010-88819475
传　　真：010-88819475
网　　址：http://vep.fltrp.com
电子信箱：vep@fltrp.com
购书电话：010-88819928/9929/9930（邮购部）
购书传真：010-88819428（邮购部）

购书咨询：（010）88819926　电子邮箱：club@fltrp.com
外研书店：https://waiyants.tmall.com
凡印刷、装订质量问题，请联系我社印制部
联系电话：（010）61207896　电子邮箱：zhijian@fltrp.com
凡侵权、盗版书籍线索，请联系我社法律事务部
举报电话：（010）88817519　电子邮箱：banquan@fltrp.com
物料号：322840001

前 言

高等职业教育的办学方针是"以服务为宗旨，以就业为导向"，培养面向生产、建设、服务和管理一线所需要的技术技能型人才。高职院校的课程必须反映职业岗位对人才的要求以及学生未来职业发展的要求，体现职业性与实践性的特点，能满足培养学生综合能力的需要。英语作为高职院校一门重要的必修课，长期以来一直被看作是孤立的公共基础课程，所教授的内容未能与学生未来的职业有效结合，很难满足不同工作岗位的实际需要。这一现状与培养目标之间的差距对新时期的高职公共英语教学提出了新的课题和新的要求，高职公共英语教学改革势在必行。

我们通过广泛调研与充分论证，在深入了解社会单位用人要求和各学校教学需求的基础上，精心策划并开发了"新职业英语"系列教材。"新职业英语"系列教材是针对高职院校公共英语课程开发的全新英语教材，以"工学结合、能力为本"的职业教育理念为指导，以培养学生在未来工作中所需要的英语应用能力为目标，在帮助学生打好语言基础的同时，重点提高听、说、读、写等应用能力，特别是工作过程中的英语交际能力，真正体现高职公共英语教学的职业性、实践性和实用性。

"新职业英语"系列教材于2009年陆续出现在广大高职院校的公共英语教学课堂上。在之后的几年里，编者与出版社一直关注和跟踪本套教材的使用情况，做了多次使用情况回访。大部分使用者认为本套教材编写理念新颖、结构合理、内容实用，体现了最新的《高等职业教育英语课程教学基本要求》（试行），对高职院校公共英语教学改革起到良好的支撑与辅助作用。

随着近年高职院校英语教学改革的深入发展，公共英语课程不仅要着重培养学生在职场环境下运用英语的基本能力，特别是听说能力，同时还要提高学生的综合文化素养和跨文化交际意识，培养学生的持续学习兴趣和自主学习能力。因此，众多院校在积极实施语言与职业密切融合的教学改革的过程中，不仅重视教学内容的职业性，同时提高对教学资源的人文性、思政性、生动性、普适性和课堂易操作性的要求。鉴于此，编者和出版社适时对本系列教材进行了修订。

教材结构

为满足基础英语与相关职业英语学习的需要，实现基础阶段与行业阶段的有机衔接，同时兼顾素质教育和个性需求，"新职业英语"系列教材分为基础篇、行业篇和素质篇三个部分。各组成部分的结构和关系如下图所示：

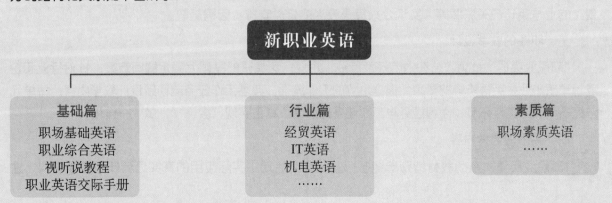

基础篇	行业篇
涵盖不同职业涉外工作中共性的典型英语交际任务，以典型工作活动中需要的英语知识和技能为线索组织教学内容，培养学生职业英语应用能力，并为其进一步学习英语打好语言基础。包括《职场基础英语》、《职业综合英语》（共两册）、《视听说教程》（共两册）和《职业英语交际手册》。其中《职业英语交际手册》是口语专项训练用书，突出口头交际能力的培养。	立足于高职院校各专业群所面向的行业，依据企业的工作流程、典型工作环节或场景设计教学内容，力求使学生具备在本行业领域内运用英语进行基本交流的能力，包括《机电英语》、《IT英语》、《经贸英语》、《医护英语》、《汽车英语》、《艺术设计英语》、《包装印刷英语》、《土建英语》、《化生英语》、《物流英语》、《市场营销英语》、《旅游英语》、《轨道交通英语》等。

素质篇

旨在提高高职学生的综合素质，兼顾学生社会发展的需求和个性发展的需要，从而实现其全面发展。包括英语技能类、英语文化类、英语应用类等。可在基础英语教学阶段和行业英语教学阶段供感兴趣的学生选用，也可在之后的提高阶段供与英语联系紧密的专业的学生选用。

为确保教材的针对性、实用性与够用度，"新职业英语"系列教材的内容均通过对各行业及职业岗位的深入调研与分析确定。基础篇与行业篇主要供高职院校英语课程必修阶段教学使用，素质篇主要供高职院校英语课程提高阶段或选修课使用。各高职院校也可根据自身的实际情况灵活安排，选择使用。

教材特色

"新职业英语"系列教材是一套顺应高职院校公共英语教学改革发展趋势、真正体现职业英语教学理念的教材，主要具有以下几方面的特点：

一、创新的教学理念

"新职业英语"系列教材以"工学结合、能力为本"的教育理念为指导，将语言学习与职业技能培养有机融合，确保教学内容与教学过程真正体现职业性与应用性，提高学生的英语交际能力与综合职业素质，从而提升他们的就业能力。

二、完备的教学体系

"新职业英语"系列教材根据高职院校公共英语基础阶段与提高阶段的教学需求，包含基础篇、行业篇和素质篇三个模块，既循序渐进、层层递进，又相互协调、相得益彰，构成了一个系统、完备的高职公共英语教学体系。不同层次、不同类别的学校，可根据地域差别、行业异同、个性需要、专业与英语的关联度等，实现公共英语教学的分类安排、因需施教。

三、职业的教学设计

"新职业英语"系列教材在对院校及行业、企业广泛调研的基础上确定编写方案，针对行业和企业对高职毕业生英语技能的要求，根据企业的工作流程、典型工作任务或场景设计教学内容。每单元浓缩一个典型工作环节，学习任务与工作任务相协调，真正实现"教、学、做"一体化。

四、实用的选材内容

"新职业英语"系列教材特别选择各行业和职业活动中实际应用的真实语料作为教学材料，注重时代性、信息性与实用性，既适用于提高语言能力，又有利于培养学生的职业素质与技能。来自于现实工作中的真实选材能为学生营造真实的语境，并通过学习内容与将来工作内容的结合提高学生的兴趣。

五、科学的测评手段

"新职业英语"系列教材采用形成性评估和终结性评估相结合的评价方法，着重考查学生的英语综合应用能力，培养学生的自主学习策略。本系列教材将提供专门的《形成性评估手册》及许多经过教学检验的形成性评估手段，既能引导学生不断进步，也不会增加教师负担。

六、立体化的教学资源

"新职业英语"系列教材根据各教学环节的需要，配备教师用书（电子版）、助教课件、评估试卷、示范课和二维码资源，提供合理的教学建议与丰富的视频等辅助资源，以方便教师备课与授课，促进教师与学生之间的互动与交流。

编写队伍

"新职业英语"系列教材由外语教学与研究出版社与徐小贞教授带领的富有教材编写经验的教师团队共同策划开发。各分册在对不同行业特点与需求以及高职院校教学情况等调研的基础上，由各行业领域中著名本科院校及高职院校的英语教师、专业教师及企业人员共商方案、合作编写。

编写说明

《新职业英语机电英语（第三版）》是"新职业英语"系列教材行业篇中机电行业的主干教材，可供机械制造类、机电设备类、自动化类等专业使用。本教材共八个单元，涵盖了机电行业最典型的工作过程，基本可以满足机电行业从业人员工作过程中的英语交际需求，也为学生将来因职业发展而进一步学习专业英语打好基础。

教材设计

一、内容组织

本教材和传统专业英语教材在内容组织上有本质的区别，不是以机械零件、金属成型、计算机辅助设计与制造、集成电路、工业机器人等学科知识的结构来编排的，而是以机电行业主要职业的典型工作过程来组织内容，即接洽客户、市场调研、机电产品设计、机电产品生产、机电产品质检、机电设备维修、机电行业生产管理、销售与售后服务。围绕着基本工作过程中的主要工作任务所需要的英语知识和技能设计英语学习任务。

二、开发模式

本教材采用"英语教师＋专业教师＋行业人员"的开发模式。由三方人员组成本教材"双师结构"的开发团队，以专业教师和行业人员为主，在对企业深入调研的基础上，分析确定机电设备等行业典型的工作过程和基本的工作任务，以英语教师为主设计语言学习任务，充分发挥各方专长。

三、适用对象

本教材以典型工作过程来组织内容，因此可以供机械制造类、机电设备类、自动化类等专业学生使用。虽然各具体职业在学科知识上可能有较大差别，但在基本要素、基本能力方面是相通的，所需要的英语知识和技能是相似的。为了工作过程的完整和选材的方便，本教材以目前高职毕业生从业最多的计算机辅助设计与制造、数控技术、模具设计与制造、机械设计与制造、机电一体化、机电设备维修和机电设备销售为基本的职业背景。

教学设计

一、兼顾语言技能的训练和语言知识的巩固

本教材应用职业教育的理论来组织内容，但最终是以语言教材的形式呈现在使用者面前，可以采用纯语言教材的方式来使用，教授本课程的教师基本不需要太多专业知识。各单元都包含听、说、读、写、译五种基本的语言技能训练和词汇、语法等语言知识的巩固。语言技能以读→听→说

→读→写→说的顺序排列，保证课堂教学中各技能的训练交叉进行，避免因相似技能训练时间太长而枯燥，同时也符合语言习得输入、输出的相关理论。语言知识部分一方面是完成语言技能培养的需要，另一方面主要是针对本阶段高职学生在英语知识方面的难点、弱点而设计，不强求知识的系统性，追求能解决实际问题。学生会深切感受所学内容与他们将来的工作紧密相关。

二、兼顾课内和课外两个教学环节

其一，本教材所倡导的教学目标之一是英语应用能力的提高，而应用能力的提高需要足够的课堂时间来组织相关教学活动，可能导致没有足够的时间来处理语言知识，这部分内容只能放在课外处理；其二，行业英语一般开设在基础英语（或职场英语）之后，基本的语言知识已不是本阶段学习的重点；其三，通过教师在课堂上布置作业引导学生在课外自学，有助于培养学生自主学习能力。课内语言技能训练和课外语言知识巩固的区分当然不是绝对的，使用者可根据教学的实际情况灵活处理，本教材模块化的布局可灵活满足个性化教学的需要。

三、兼顾语用意识、跨文化意识、学习策略意识和职业技能

上述技能的培养以隐性、非系统的方式分散在各单元中，旨在使学生在掌握相关英语知识和英语技能之外，能产生一些语用意识、交际意识和学习策略意识，并最终形成英语的职业技能，包括专业能力、方法能力和社会能力。教师在上课过程中遇到相关内容可灵活处理：比如对基础一般的学生可略去，对基础较好的学生则可充分讲解甚至适当补充，使他们具备更完善的职业英语综合能力，在未来的职场中更有竞争优势。

四、兼顾优秀的教学理论和教学方法

职业性和实践性是本教材的区别性特征，同时也继承了外语教学及教材编写一些优秀的理论和做法。在教学方法上，不局限于某一特定理论而是博采众长、为我所用，强调听、说、读、写、译各种技能的整体培养，教学以学生为中心，以任务、项目的形式让学生在做中学，真正实现教学做一体化，使学生最终获得综合能力的提高。

教材结构

本书每单元为一个典型工作任务，八个单元组成机电设备等行业主要职业最典型的工作过程，同时每单元又由不同的微任务组成，这些微任务共同组成一个完整的微工作过程。每单元均分为课内、课外两个环节，课内环节包括Unit Objectives, Warming-up, Reading A, Listening, Speaking, Reading B, Writing, Project, Self-evaluation九个部分，课外环节包括New Words and Expressions, Vocabulary and Structure, Grammar和Fun Time四个部分。每单元8-10个学时，全书计划用时64-80学时，可供行业英语教学阶段1-2学期（72学时或36学时）使用。

本教材对全书的难度和梯度进行了合理的控制，Reading A的长度控制在350—400词，Reading B的长度控制在250词左右。

一、课内部分

1. Unit Objectives

单元目标部分，提供单元所涉及的主要职业技能，在学习开始之前对单元的职业技能目标有一个清晰的了解。

2. Warming-up

单元主题的导入部分，设计一些与单元主要工作环节相关的简单有趣、操作性强的活动，既能激发学生的兴趣，导入对主题的学习，又能让学生就此话题交流自身的知识与生活经验，为后面的学习活动做好准备。

3. Reading A

阅读部分，围绕一篇阅读材料展开一系列的活动。每单元根据不同职业岗位或工作活动所涉及的专业知识选材，提供机电方面的相关信息，侧重阅读理解和语言知识的输入。

4. Listening

听力部分，围绕工作过程所涉及到的典型职业活动场景提供相应的听力练习，形式多样，活泼有趣，旨在让学生听懂职场日常涉外业务活动中基本的表达。建议教师循序渐进地引导学生完成听力任务。

5. Speaking

口语活动部分，和听力部分所涉及的交际功能基本对应，强调语言的输出，在熟悉典型交际情景的同时，比较熟练地掌握相关交际功能的语言表达方式，任务形式包括对话、小组活动、角色表演、个人陈述等。

6. Reading B

实用阅读部分，选取职场工作中的实用题材和文体，例如机电行业企业业务关系的建立、调研设计、产品每日生产报表、质检表格等，丰富了本单元学习内容的同时，重在培养学生把握真实工作语料的能力。

7. Writing

实用写作部分，以任务的形式、围绕相关工作过程所需要的应用文设计写作练习，例如业务信函、调研分析报告、生产报表和库存报表、生产技术报告、机电设备展会邀请函等，使学生掌握机电行业中实际应用的文体写作。

8. Project

单元内容的应用部分，以项目的形式让学生实践单元的语言技能和职业技能，同时复习单元所涉及的主要微工作过程，从而把英语和职业联系起来，把学习和工作联系起来。项目内容与学生日常生活及将来工作密切相关，提供明确的操作指令，一般以小组的形式完成。

9. Self-evaluation

学生自我评估部分，与单元目标相呼应，从语言学习的角度引导学生自行检查学习效果，进而培养学生拥有一定的英语学习策略和自主学习能力。

二、课外部分

1. New Words and Expressions

Reading A 和 Reading B 的生词和短语部分，包括生词、短语、专有名词等，提供音标、词性和词义。本部分可让学生课外自学，也可根据需要在课堂上处理。

2. Vocabulary and Structure

Reading A 和 Reading B 的词汇和结构练习部分，包括拼写、词形变换、词义运用、结构和重点表达等练习。可以在课堂上与 Reading A、Reading B 一起处理，也可用作课外作业。

3. Grammar

语法部分，总结学生在语法上的难点或重点，侧重语法知识在交际活动中的应用和通过应用领悟语法规则。分为三个模块：模块一通过语言活动，使学生对某一语法产生感性认识；模块二是对该语法提纲挈领的讲解；模块三是对该语法在交际活动中的应用。本部分自成体系，其中第一个模块和第二个模块可在课堂上处理，第三个模块留给学生课外完成。

4. Fun Time

此部分涉及英语语言趣味学习、小幽默等，以开拓学生视野，启发学生思维，增长学生经验。

另外，为方便教师实施教学，本教材还配有相应的教师用书和助教课件，教师可以从外研社高等英语教学网（heep.unipus.cn）上下载。

编写队伍

"新职业英语"系列教材总主编为徐小贞教授。《新职业英语机电英语（第三版）》主编为胡庭山，编者为金曙、吴敏、蔡嵘、杨茂霞、徐晓燕。

音频资源

《新职业英语机电英语（第三版）》采用了二维码扫描阅读技术。使用方法：首先扫描右侧二维码，下载外研随身学app（职教版）。注册后通过app的扫码功能，扫描教材中的二维码，即可获取配套的音频资源。

Contents

Unit	Reading A	Listening and Speaking	Reading B	Writing	Project	Vocabulary	Grammar	Fun Time
Unit 1 Establishing Business Relations (p1)	Create Excellence—Blue-sky Mold	Meeting People and Taking Orders	Letters of Establishing Business Relations	Letters of Establishing Business Relations	Establishing Business Relations	New Words and Expressions Vocabulary and Structure	Part of Speech and Sentence Components	How to Tell a Joke
Unit 2 Market Research (p25)	Gas Compressor Market in China—The Past and the Future	Doing Market Research	Expert Interview—MEMS Switch Applications for Semiconductor Test Market	Mini-report on Market	Doing Market Research	New Words and Expressions Vocabulary and Structure	Modal Verbs	Jokes
Unit 3 Product Designing (p51)	Computer-aided Design	Discussing About Product Design	Gear Reducer—A Great Design	A Design Scheme	Designing an Electric Bike	New Words and Expressions Vocabulary and Structure	Tenses	The Funny English Language
Unit 4 Production (p75)	Computer-aided Manufacturing	Discussing About Product Manufacturing	Plastic Molding	Daily Production Report	Manufacturing a Product	New Words and Expressions Vocabulary and Structure	Sentences	Wrong Email Address
Unit 5 Product Inspection (p97)	Inspection and Testing	Discussing About Inspections	Inspection Record of a Precision Bench Lathe	Fax & Notice	Making a Final Inspection	New Words and Expressions Vocabulary and Structure	Attributive Clauses	A Humor
Unit 6 Installation and Maintenance (p121)	Mounting the Lathe to a Board	Dealing with the Repairing	Cleaning the Lathe	Cleaning Instructions & A Follow-up Letter	Learning About the Preventive Maintenance of Machines	New Words and Expressions Vocabulary and Structure	Non-finite Verbs	A Humor
Unit 7 Operational & Technical Management (p145)	PDM System and Mold Design	Technology Innovation & Warehouse Management	Responsibilities of the Warehouse Worker and the Warehouse Manager	Certificate of Warehousing by the Consignee	Developing and Improving a Product	New Words and Expressions Vocabulary and Structure	Subjunctive Mood	A Humor
Unit 8 Marketing and After-sales Service (p169)	Doosan Infracore's Marketing Strategy in China	Selling a Product & Dealing with Customers' Complaints	Invitation to India Machine Tools Show	A Reply to Invitation to an Exhibition	Marketing the Products	New Words and Expressions Vocabulary and Structure	Noun Clauses	A Humor

UNIT 1

Establishing Business Relations

Unit Objectives

After studying this unit, you are able to:
- demonstrate the strengths of your company
- arrange a meeting and receive clients
- negotiate the price with clients
- build business relations with a company
- understand a contract

English for Mechanical & Electrical Engineering

Warming-up

Task 1 As a marketing assistant, Bruce may experience the following situations. Match each situation with its corresponding picture.

☐ Meeting a client.

☐ Signing a contract.

☐ Providing necessary information for the manager.

☐ Briefing a product.

Task 2 The following things are what Bruce usually does for taking orders. Arrange them in the order of time and explain.

a. Show the product line to the client.
b. Make an inquiry, offer and counter-offer.
c. Make an appointment with the client.
d. Sign the contract.
e. Introduce his company to the client.

❶___ > ❷___ > ❸___ > ❹___ > ❺___

Reading A

Task 1 Before reading the passage, try to answer the questions about company profile.

1. What may be included when introducing a company to the client?
2. Which company's introduction or ads ever impressed you most? And why?

Create Excellence — Blue-sky Mold

Blue-sky Mold is one of the most well-known extrusion mold manufacturers in China. We own two subsidiary companies: TA Mold and TC Mold and four mold classifications and produce hundreds of different products.

The original company, TA Mold, was built in 1993. The 30-acre plant is located in Ningbo City, Zhejiang Province, a city famous as the "hometown of molds" and "kingdom of plastic". In 2007, we expanded our business by setting up a new company called TC Mold.

An experienced professional extrusion mold manufacturer, we have created our own unique extrusion integrated system. We are leaders in the field of mold design, thermoplastic fine inching control, PVC low foam technology, WPC raw material formulas and extrusion operating techniques.

We have developed many products that are widely used in the construction industry, interior and external decoration and packaging as well as many of life's daily necessities.

We are committed to excellence in all aspects of our business. To this end, we aim to work closely with our customers to ensure that their demands are met and that our products are competitively priced without compromising on quality.

Our company is proud to offer expert

support and advice to our customers to enable them to select or develop the most effective molded products for their requirements. Our technology support teams provide experienced advice and support at every stage of the business and we are also able to provide on-the-spot training to enable customers to use our products most effectively. Nonetheless, we are not complacent and constantly seek to improve the quality of our service and our products.

Because of our rich experience, advanced equipment and effective production processes, our products are now exported to over 40 countries in Europe, America, Southeast and Central Asia. We aim to provide all our customers with the very best Chinese extrusion mold products and technological support.

We look forward to cooperating with you to create a bright future.

English for Mechanical & Electrical Engineering

Task 2 Read the passage and pick out the areas mentioned that the extrusion molds can be used in.

medical industry	( )	architecture material	( )
construction industry	( )	external decoration	( )
sports equipment	( )	packaging	( )
daily necessities	( )	IT industry	( )
interior decoration	( )	gardening	( )

Task 3 Read the passage again and fill in the blanks.

1. Two subsidiary companies: _____
2. The location of the original company: _____
3. The field the company specializes in: _____
4. The fields in which the company is in the leading level: _____
5. What support the company offers: _____
6. The countries or areas the products are exported to: _____
7. The aim the company strives for: _____
8. The wishes of the company: _____

Task 4 While introducing a mold company, what else do you think the customers may be interested in? Discuss with your classmates.

4

Unit 1 Establishing Business Relations

Listening

Task 1 Listen to the conversation and match the people with the correct information.

Martin Jones

Mr. Li

Wang Ning

- marketing assistant of TA Mold
- a prospective customer from Australia
- marketing manager of TA Mold
- works in Leigh Mardon Company
- wants to get more information about the products

Task 2 Listen to the conversation between Martin Jones and Wang Ning and fill in the blanks with what you hear.

Wang Ning: Hello! This is Wang Ning, marketing assistant of TA Mold.

Mr. Jones: Hello, Miss Wang. This is Martin Jones from Leigh Mardon Company. I'm interested in your 1 _____ mold. I've seen some samples in Guangzhou Fair. They are very 2 _____ .

Wang Ning: Thank you for your interest in our products. They are 3 _____ products and have been 4 _____ to more than 40 countries and areas with high identification.

Mr. Jones: I see. And could you give me a 5 _____ introduction of your products?

Wang Ning: Okay, Mr. Jones. But you know, for each model, we have a 6 _____ of products. May I fax you a catalog?

Mr. Jones: No, you needn't. I have got one copy. I'm really 7 _____ in the prices of your products.

Wang Ning: In that case, I think we do need a 8 _____ .

Mr. Jones: That's 9 _____ what I'm thinking. When will it be convenient for you?

Wang Ning: It is up to you. But I hope you can come to visit our company.

Mr. Jones: I'll fly to Ningbo next Monday morning. Is that alright for you?

Wang Ning: That's great! I will meet you at the 10 _____ .

Mr. Jones: Thank you. I will email you the flight number. See you then.

Wang Ning: See you.

English for Mechanical & Electrical Engineering

Task 3 Martin Jones has to attend an important conference next Monday morning and he calls Wang Ning to rearrange the time. Listen to the conversation and choose the best answer to each question.

1. When should Mr. Jones fly to Ningbo according to the original plan?
   A. This Monday morning.   B. Next Monday morning.
   C. This Thursday morning.   D. Next Thursday morning.
2. Why does Mr. Jones want to change the time?
   A. Because he will be in the US on business that time.
   B. Because he does not want to meet Wang Ning.
   C. Because he will have a meeting to attend.
   D. Because he has already met Wang Ning.
3. When will Mr. Jones fly to Ningbo?
   A. This Monday morning.   B. Next Monday morning.
   C. Next Thursday morning.   D. Monday morning the week after the next.
4. What will Miss Wang do next Thursday morning?
   A. Meet Mr. Jones.   B. Be on business.
   C. Attend a meeting.   D. Be on vacation.
5. What can we infer from the conversation?
   A. Mr. Jones often puts off his appointments with others.
   B. Miss Wang often flies to the US on business.
   C. Miss Wang and Mr. Jones have not reached an agreement on when they will meet.
   D. Face-to-face talk is sometimes a must in the business field.

Task 4 Wang Ning meets Martin Jones at the airport. Listen to the conversation and answer the following questions.

1. Where did Mr. Jones get the information about Ningbo city?
   _____
2. What impression did Mr. Jones have about Ningbo city?
   _____
3. Has Mr. Jones ever visited China? How about Ningbo city?
   _____
4. Where will Mr. Jones stay?
   _____
5. When will the manager meet Mr. Jones?
   _____

Unit 1 Establishing Business Relations

Task 5 Mr. Jones visits TA Mold and feels quite satisfied. Now he is sitting at the negotiating table and talking about the prices with Mr. Li. Listen to the conversation and complete the sentences with the information you hear.

Mr. Jones: OK. Mr. Li. I'm much impressed by your working environment and the workers' efficiency here.

Mr. Li: Thank you. In fact, we have 1 _____, and our products have been exported to over 40 countries and areas.

Mr. Jones: What services will you provide your customers with?

Mr. Li: We could 2 _____, selection of extruder machine, mold technique and operation training on the spot.

Mr. Jones: Great! We are very interested in your plastic molds. I'd like to know your latest prices for plastic molds, FOB(Free on Board 船上交货价格) Ningbo.

Mr. Li: Due to the increase of international demands on plastics, the prices of the plastic molds 3 _____ this month. But we can offer you some discount, if you place a big order.

Mr. Jones: Oh, you mean the prices are flexible?

Mr. Li: Yes, especially 4 _____.

Mr. Jones: Well, what discount can I get?

Mr. Li: Three percent off! That is the maximum we can offer.

Mr. Jones: Mr. Li, as you know, 5 _____. It is easy for us to get the same molds at much lower prices.

Mr. Li: We are in the leading place in the plastic molds in the world market. Our products are competitive in quality and prices as well.

Mr. Jones: Oh, yes. But ah... Anyway, it's over and above what we expected.

Mr. Li: Okay. 6 _____ for you?

Mr. Jones: Five percent off?

Mr. Li: Five percent off. Well... that is too much. How about four percent off?

Mr. Jones: Okay. It seems that I have no other choice but to accept it.

Mr. Li: I'm glad that we've settled on the price.

English for Mechanical & Electrical Engineering

Speaking

Task 1 Work in pairs. Practice making short phone calls with the words provided according to the example below.

> Example: Tom Green/Yohama Co., Ltd./Bruce Bowen/call back
>
> **A:** TA Industrial Design Company. Can I help you?
>
> **B:** Yes, this is Tom Green from Yohama Co. Ltd. and I'd like to speak to Bruce Bowen, please.
>
> **A:** Just a minute and I'll put you through... I'm sorry, he's having a meeting right now. Can I get your number and ask him to call you back?
>
> **B:** Uh, 408-816-008. I am available from 9 to 11 am. I would appreciate it if he could call back soon.
>
> **A:** Don't worry. He will, Mr. Green.
>
> **B:** Thank you.

1. Brian Reeves/General Motors/Zhang Xinmin/call back
2. He Haiyang/Zhejiang Import & Export Co., Ltd./Joe Feigan/fax the documents
3. Tom Black/Nissan/Wang Peng/send a price list

Task 2 Work in pairs. Suppose the manager of Yohama Co., Ltd. in Japan is interested in the products of TA Industrial Design Company. Practice making a phone conversation. You may use the phrases or expressions listed below.

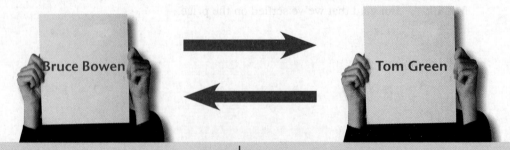

Bruce Bowen	Tom Green
Can I help you?	I am interested in...
Have a face-to-face talk.	Would you please...
When will it be convenient for you?	Be free after 5:30 pm.
Where to meet?	The City Hotel, Ningbo.

Unit 1 Establishing Business Relations

Task 3 Work in pairs. Bruce Bowen meets Tom Green at the City Hotel. Play their roles according to the instructions below.

Bruce Bowen

Introduce the company briefly.

Tell the price.

Show the competitiveness of the products.

Agree to offer a discount.

Tom Green

Show interest in the WPC molds.

Ask for the price.

Think the price is too high.

Agree.

Task 4 Work in pairs. Assume you and your partner are the seller and buyer of the smart phone respectively. Make a conversation and act it out. Some useful sentences are provided below for your reference.

❖ **Requesting for discounts**

Could you make it a little cheaper?
Is it possible for you to reduce your price by 10%?
A deal is possible if you reduce your price.

❖ **Bargaining**

You are asking too much for this product.
The price is too high. I can buy the same thing from ABC Company at one third of your price.
If you order a larger amount, we can reduce the price a little.
Usually, we offer no discount for it.

❖ **Striking a deal**

Let's meet each other half way.
OK, let's call it a deal!
All right. I now agree with you on that.

Reading B

Letters of Establishing Business Relations

Establishing business relations is the first step to develop trade ties. Since business growth and expansion largely depend on the establishment of business relations, writing appropriate business letters in this respect is vitally important.

When writing a letter to start business with another company, you are supposed to tell your readers how you get this address and what your business line is, then state your purpose and request, and finally express your sincere wish to cooperate in future.

Here is a sample letter.

Dear Ms. Wang Ning,

We should like you to send us a catalog and price list regarding your plastic mold products and also samples of the different item models. You were recommended by Mr. Thompson, your business representative here, who convinced us that you are among the long-standing mold companies with high reliability in your country.

Our company is engaged in both civil and industrial construction, and we are interested in purchasing a number of products from your company. As we are about to invest more on the civil construction, it is essential that our suppliers be both competitive in terms of price, and extremely reliable.

If your price and quality are attractive, you may expect a big order from us.

Our opening bank is National Australia Bank Limited (NAB), Head Office, Sydney, from whom you can have any reference regarding our financial standing.

We are looking forward to hearing from you very soon.

Yours Sincerely,
Martin Jones
General Manager of Leigh Mardon Company in Australia

If you reply to a letter of this kind, you should try to answer all the questions with the necessary information the other party required. Please remember to reply politely even if you are unable to meet the needs. The reasons should be made clear so as to leave space for future cooperation.

Unit 1 Establishing Business Relations

Task 1 Read the passage and decide whether the following statements are true (T) or false (F).

☐ 1. Establishing business relations is the first step to expand trade.
☐ 2. In international trade, one may establish business relations with the companies in other countries through local business representatives.
☐ 3. Writing an appropriate business letter is necessary to establish business relations.
☐ 4. When writing a letter to start business with another company, you should tell your readers how you get to know them.
☐ 5. When writing a letter to start business with another company, you needn't have to express your hope to cooperate with them.
☐ 6. You don't need to answer all the questions the other party required when you reply to the letter.

Task 2 Read the passage again and answer the following questions.

1. Why does Martin Jones write this letter?

2. How does Martin Jones get to know Ms. Wang's company?

3. Why does Leigh Mardon Company want to establish business relations with Blue-sky Mold?

4. What business is Leigh Mardon Company engaged in?

5. Why does Martin Jones refer to NAB?

Task 3 Translate the following paragraph into Chinese.

> To give you a general idea of our products, we enclose a complete set of leaflets showing various products being handled by this company with detailed specifications and means of packing. The price list and samples will be sent upon receipt of your specific inquiries.

Writing

Task 1 Wang Ning wrote a letter in reply to Martin Jones. Please finish the letter according to the letter in Reading B.

Dear Mr. Martin Jones,

Your May 16 letter has been received. We are very glad to find that your company _____.

Our company was founded in 1993, and ever since then we have gradually built our fame in China as well as in the world. We specialize in _____, and we are in the leading place in _____.
We can assure you that _____. Furthermore, we promise to provide you _____.

The discount you care about depends on _____, which needs some further discussion.

Enclosed we are sending you _____.

Finally, we would like to extend our sincere invitation to you, hoping _____.
_____.

Yours faithfully,
Wang Ning
The Marketing Assistant from Blue-Sky Mold in China

Task 2 The following is a delivery contract sample. Fill in the blanks in the contract according to the information in this unit.

Contract No.: PM00347
Signed at: City Hotel, Ningbo, China
Date: July 28, 2020
The Buyer: 1 _____
The Seller: 2 _____

The Buyer agrees to buy and the Seller agrees to sell the following goods on terms and conditions as set forth below:
Name of Commodity, Specifications: YL26, Made in China
Quantity: 150 Units

Unit Price: $100
Total Value: 3 _____
Time of Shipment: Aug 28, 2020
Port of Loading: 4 _____
Port of Destination: Melbourne Port
Terms of Payment: By 100% Confirmed, Irrevocable and Sight Letter of Credit to remain valid until the 15th day after shipment.

Project

Project Guidelines

This project aims to go through the whole process of establishing business relations. The task is divided into three steps. Step One is about the introduction of the company. Step Two focuses on the client reception. Step Three rests on the process of business negotiation between the company and the client.

Please follow the *Task Description* to complete the project.

Task Description

Step One
- Organize a small group of 4-6 students in your class;
- Share the work of researching online resources for information about a company (e.g. company profile, products and services, contact information, etc.);
- Give a presentation to introduce your company and products with the help of the online resources you found.

Step Two
- Discuss with your partners what you should do during the reception;
- Talk about what you should pay attention to in doing the reception work and write down the tips;
- Share the discussion results with your classmates;
- Divide your group into two sides: one side being the company staff, the other being the potential customers;
- Take turns to play each role at the first meeting.

Step Three
- Divide your group into two and discuss in the sub-groups how to negotiate the price;
- Take turns to role-play the situation of negotiating the price;
- Search online for videos about signing a contract and then practice the ceremony of signing the contract.

English for Mechanical & Electrical Engineering

⚙ Self-evaluation

Rate your progress in this unit.	D	M	P	F*
I can demonstrate the strengths of our company.	☐	☐	☐	☐
I can communicate with potential clients.	☐	☐	☐	☐
I can arrange a meeting and receive clients.	☐	☐	☐	☐
I can negotiate the price with clients.	☐	☐	☐	☐
I can establish business relations with a company.	☐	☐	☐	☑
I can understand a contract.	☐	☑	☐	☐

*__Note__: Distinction, Merit, Pass, Fail

New Words and Expressions

Reading A

New Words

classification /ˌklæsəfəˈkeʃən/ n. 类别，种类
competitively /kəmˈpɛtətɪvlɪ/ ad. 有竞争力地
complacent /kəmˈplesnt/ a. 自满的，自鸣得意的
compromise /ˈkɑmprəˌmaɪz/ n./v. 妥协，让步
constantly /ˈkɑnstəntlɪ/ ad. 不断地，经常地
construction /kənˈstrʌkʃən/ n. 建筑 (物)
cooperate /koˈɑpəˌret/ v. 合作
decoration /ˌdɛkəˈreʃən/ n. 装饰，装饰品
ensure /ɪnˈʃʊr/ v. 确保，保证
expand /ɪkˈspænd/ v. 扩大，扩展
export /ˈɛkspɔrt/ v. 出口
external /ɪkˈstɜnl/ a. 外部的，外面的
formula /ˈfɔrmjələ/ n. 配方，公式
integrated /ˈɪntəˌgretɪd/ a. 集成的，综合的
manufacturer /ˌmænjəˈfæktʃərɚ/ n. 生产商，制造商
original /əˈrɪdʒənl/ a. 最初的，最早的
process /ˈprɑsɛs/ n. 程序，流程
professional /prəˈfɛʃənl/ a. 专业的，职业的
reference /ˈrɛfərəns/ n. 证明文书
spot /spɑt/ n. 现场
subsidiary /səbˈsɪdɪˌɛrɪ/ a. 附属的，副的
thermoplastic /ˌθɜməˈplæstɪk/ n. 热塑性塑料
unique /juˈnik/ a. 独一无二的，独特的

Phrases & Expressions

be committed to 致力于
in the field of 在……领域
cooperate with 与……合作，与……共同努力

Technical Terms

extrusion mold 挤压式模具
fine inching control 精密缓动控制
PVC low foam technology PVC (Polyvinyl Chloride 聚氯乙烯) 低发泡技术
WPC raw material formula WPC (木塑复合) 原材料配方

Reading B

New Words

appropriate /əˈproprɪɪt/ a. 适当的，恰当的
catalog /ˈkætlˌɔg/ n. 目录
civil /ˈsɪvl/ a. 民用的
convince /kənˈvɪns/ v. 使确信，使信服
essential /ɪˈsɛnʃəl/ a. 基本的
item /ˈaɪtəm/ n. 项目，条款
long-standing /ˌlɔŋˈstændɪŋ/ a. 久经考验的，经久不衰的
model /ˈmɑdl/ n. 型号，样式
party /ˈpɑrtɪ/ n. 一方，当事人
purchase /ˈpɜtʃəs/ n./v. 购买
recommend /ˌrɛkəˈmɛnd/ v. 推荐
reliability /rɪˌlaɪəˈbɪlətɪ/ n. 可靠 (reliable 是其形容词形式)
representative /ˌrɛprɪˈzɛntətɪv/ n. 代表
sample /ˈsæmpl/ n. 范例，样品
sincere /sɪnˈsɪr/ a. 真诚的
supplier /səˈplaɪɚ/ n. 供应商，厂商
vitally /ˈvaɪtlɪ/ ad. 极为，生死攸关地

Phrases & Expressions

be about to 将要，正打算
be engaged in 专营于
head office 总部
in terms of 根据，按照，用……的话，在……方面
in this respect 在这方面
invest... on 投资于
prior to 在前，居先

Proper Names

National Australia Bank (NAB) 澳大利亚国立银行
Sydney 悉尼

Vocabulary and Structure

Task 1   Write out the words in Reading A according to their meanings in the right column. The first letters are already given.

e_____	to make certain that something will happen properly
c_____	determined or trying very hard to be more successful than other people
e_____	to sell or transfer abroad
p_____	of or belonging to a profession
e_____	to become larger in size, number or amount
d_____	things used for beautifying something
c_____	all the time or very often
c_____	to make a concession

Task 2   Fill in each blank with the appropriate form of the word given in the brackets.

1. This model may be (classification) _____ as either an optimizing model or simulation model.

2. My company (manufacture) _____ toys and games for children.

3. "Have you consulted any other member of my (professional) _____?" he asked.

4. Our foreign trade (expand) _____ during recent years.

5. We must hurry back to the Armory and get the (decorate) _____ done.

6. I (commit) _____ to attending the farewell dinner.

7. I wanted to go to Greece, and my wife wanted to go to Spain, so we (compromise) _____ on Italy.

8. Despite yesterday's win, there is clearly no room for (complacent) _____ if the team wants to stay top of the league.

9. Miller was a quiet and (reliability) _____ man.

10. Regular exercise is (vitally) _____ for your health.

Task 3 Complete the following sentences with the words or phrases given below. Change the form if necessary.

| set up | expand | in the field of | cooperate | be committed to |
| look forward to | so as to | convince | be engaged in | invest |

1. The church seeks to _____ closely with local schools.
2. You must always create things to _____, to enjoy and you must always reward yourself each week.
3. She _____ a new world record in the 100 metres.
4. He ran quickly _____ catch the last bus.
5. The doctor is _____ that she does not need to lose weight.
6. He does wholesale business, while his brother _____ retail business.
7. People _____ education worked especially hard.
8. He _____ his short story into a novel.
9. He _____ his savings in a business enterprise.
10. For all these years, she has _____ the cause of women's liberation.

Task 4 Make sentences with the same pattern as is shown in the examples.

Example 1: we/have developed many products that are widely used in the construction industry, interior and external decoration and packaging/many of life's daily necessities
— We have developed many products that are widely used in the construction industry, interior and external decoration and packaging *as well as* many of life's daily necessities.

Example 2: she/has got a boy/girl
— She has got a boy *as well as* a girl.

1. Harry/has knowledge/experience

2. he/grows flowers/vegetables

3. hatred/is blind/love

4. he/drove carefully/slowly

5. air/water/is needed to make plants grow

> *Example 1:* our suppliers/both competitive/in terms of price/and/extremely reliable
> — *It is essential that* our suppliers be both competitive in terms of price, and extremely reliable.
>
> *Example 2:* a student/know/something about a computer
> — *It is essential that* a student should know something about a computer.

1. land/put/under grass and corn cultivated
   _____

2. both wind strength and direction/serve our purpose
   _____

3. my judgment/balance/and/my brain/clear
   _____

4. everyone/be kept informed about/what is involved and just how the new system will impact upon them
   _____

5. in doing this kind of work/the water/clean/and/free from rust or dirt
   _____

Task 5 Translate the following sentences into English using the words or phrases given in the brackets.

1. 我们在机床 (lathe) 制造方面处于领先地位。(in the field of)
   _____

2. 我们的产品无论是价格还是质量，都在市场上具有竞争优势。(in terms of)
   _____

3. 我们和同事们协力制订生产计划。(cooperate with)
   _____

4. 我们一直致力于在产品与服务上不断取得进步。(commit to)
   _____

5. 他正要离开的时候，有人叫住了他。(be about to)
   _____

Grammar

Parts of Speech and Sentence Components

Task 1 Identify the part of speech of each underlined word in the following paragraph.

Andrew didn't go to the cinema with other students. Rachel told him they were going there, but he wanted to finish his homework. Andrew isn't very sociable. He stays in his room and concentrates totally on his studies. He's an excellent student, but he doesn't have much fun.

1. to _____prep._____
2. cinema _____
3. other _____
4. told _____
5. they _____
6. there _____
7. he _____
8. finish _____
9. sociable _____
10. in _____
11. and _____
12. totally _____
13. an _____
14. excellent _____
15. but _____
16. fun _____

Task 2 Identify the part of speech of the underlined word in each sentence.

1. We went to a wonderful show in Beijing. _____n._____
2. Jenny wanted to show Jack her photos. _____
3. Henry thought Claire looked beautiful. _____
4. A strange thought came into her head. _____
5. The windows are clean. _____
6. We should clean the windows. _____
7. Wendy is feeling quite tired now. _____
8. Studying all day had tired Wendy out. _____
9. We did some hard work. _____
10. They worked hard. _____

Task 3 Match each underlined word with its corresponding component in the sentence. Some sentence components may be used twice.

1. <u>Time</u> flies.
2. The scenery is <u>beautiful</u>.
3. My father <u>works</u> in a large company.
4. <u>She</u> loves music very much.
5. He passed me <u>a book</u>.
6. The visitor gave an apple to <u>the monkey</u>.
7. I have the <u>perfect</u> present for her.
8. He came back <u>sick</u>.
9. John runs <u>quickly</u>.
10. We <u>students</u> should study hard.

A. subject
B. predicate
C. object
D. attribute
E. adverbial
F. complement
G. predicative
H. appositive

Task 4 Fill in each blank with the proper form of the word given in the brackets.

1. Living in the country is less __expensive__ (expense) than living in the city.
2. We need to reduce our _____ (depend) on oil as a source energy.
3. The chairman emphasized his ideas by speaking more _____ (loud).
4. Some foreign businessmen in China are spending a lot of time in _____ (learn) Chinese.
5. Some people do believe that smoking will _____ (certain) cause lung cancer.
6. I'm _____ (real) sorry for the mistake our office worker made last month.
7. Her _____ (beautiful) had faded over the years.
8. The sellers allowed us to pay them on a _____ (month) basis.
9. It's _____ (danger) for women to walk alone at night.
10. I was impressed by the _____ (deep) and complexity of the book.

词类与句子成分

词 类	缩 写	中文名	例 子	概念或功能
noun	*n.*	名词	boy, milk, bike	表示人、物及抽象概念的名称,除谓语(predicate)外,可充当句子任何成分。
pronoun	*pron.*	代词	you, I, my, yours	代替名词或起名词作用的短语、句子,可充当主语(subject)、表语(predicative)和宾语(object)等。
adjective	*adj.*	形容词	good, happy, nice	描绘人或事物的特征、性质、状态,可作定语(attribute)、表语、补语(complement)等。
adverb	*adv.*	副词	well, badly, quickly	修饰动词、形容词、副词或全句,表达时间、地点、程度、方式等概念,主要用作状语(adverbial)。
verb	*v.*	动词	be, cut, run, jump	表示动作或状态,作谓语或作为系动词(linking verb)与表语连用。
numeral	*num.*	数词	two, third	表示数目多少或顺序,可作主语、宾语、定语、同位语(appositive)等。
article	*art.*	冠词	a, an, the	限定、说明名词的所指。
preposition	*prep.*	介词	at, in, with, for	表示词与词、词与句之间的关系,与介词宾语构成短语,可作表语、定语、状语、补语等。
conjunction	*conj.*	连词	and, so, because	连接词、短语或句子,表达逻辑关系。
interjection	*interj.*	感叹词	oh, yeah, ouch	表示喜怒哀乐等感情或情绪。

Comprehensive Exercises

Task 1 Cross out the wrong word in each word pair underlined.

Dear Christine,

Well, here I am in Australia. Thank you for your 1 <u>kind/~~kindly~~</u> letters. You ask me what it's like here. I must say it's pretty 2 <u>good/well</u>! The language school is very 3 <u>efficient/efficiently</u> organized. On the first morning we had to take a test, which I found rather 4 <u>hard/hardly</u>. However, I got a 5 <u>surprising/surprisingly</u> good mark, so I'm in the second class. I didn't take much at first, because I couldn't think of the words 6 <u>quick/quickly</u> enough, but 7 <u>late/lately</u> I've become much more 8 <u>fluent/fluently</u>. I'm staying with a family who live 9 <u>near/nearly</u> the school. They are quite 10 <u>pleasant/pleasantly</u> although I don't see much of them because I'm always so 11 <u>busy/busily</u> with my friends from school. I was surprised how 12 <u>easy/easily</u> I made friends here. They come from 13 <u>different/differently</u> parts of the world and we have some 14 <u>absolute/absolutely</u> fascinating discussions. I do hope you will be able to join me here next term. I'm sure we'd have 15 <u>good/well</u> fun together.

Best wishes,

Celia

Task 2 Rewrite each of the following sentences without changing its meaning.

1. The boy is a quick learner.

   The boy learns _____.

2. The man can cook really well.

   The man is a(n) _____.

3. Your behavior was quite foolish.

   You behaved _____.

4. The hotel staff treated us in a very friendly manner.

   The hotel staff were _____.

5. He proposed that we put off our meeting, which is unreasonable.

   His _____ that we put off our meeting is unreasonable.

6. Philippa is usually a hard worker.

   Philippa usually works _____.

7. Tom looked sad when he saw the injured dog.

   Tom looked _____.

8. I wish you could swim fast.

   I wish you were a(n) _____.

9. She speaks perfect English.

   She speaks English _____.

10. I didn't go out because of the heavy rain.

    I didn't go out because it rained _____.

Unit 1 Establishing Business Relations

Task 3 Identify the part of speech and sentence component of each underlined word or phrase in the following paragraph.

Mike thinks Lily is beautiful. He loves her deeply and dreams of marrying her, but
 ①   ②   ③ ④   ⑤   ⑥

unluckily he is rather old for her. Today they are at a café. With their friends Jenny and Lucy on
   ⑦   ⑧   ⑨   ⑩   ⑪   ⑫   ⑬

the spot, Mike can't get romantic with Lily. But he might buy her some flowers later to make
                                                              ⑭   ⑮

her happy.
    ⑯

Word/Phrase	Part of Speech	Sentence Component
1. ① Mike		
2. ⑨ Today		
3. ⑬ Jenny		
4. ⑮ flowers		
5. ② beautiful		
6. ⑧ old		
7. ⑭ some		
8. ⑯ happy		
9. ④ loves		
10. ⑩ are		
11. ③ He		
12. ⑥ her		
13. ⑤ deeply		
14. ⑦ unluckily		
15. ⑪ at a café		
16. ⑫ With their friends		

Task 4 Fill in each blank with the proper form of the word given in brackets.

When Helen Keller was born she was a(n) 1 _____ (health) baby. But 2 _____ (unfortunate), when she was 19 months old, she had a sudden fever. Later, the fever 3 _____ (disappearance), but she became blind and deaf.

When Helen was seven years old, a teacher, Anne Sullivan, came to live with Helen's family. First, Anne taught Helen how to talk with her fingers. Then Anne taught Helen to read by the Braille system. Helen learned these things 4 _____ (quick). However, learning to speak was harder. Anne continued to teach Helen with 5 _____ (patient). 6 _____ (final), when Helen was 10 years old, she could speak 7 _____ (clear) enough for people to understand her.

Helen went to an institute for the blind, where she did very well in her studies. Then she went to college, where she graduated with honors when she was 24 years old. Helen traveled 8 _____ (extensive) with Anne. She worked 9 _____ (tireless), traveling all over America, Europe, and Asia to raise money to build schools for blind people. Her main message was that disabled people are like everybody else. They want to live life fully and 10 _____ (normal). Helen wanted all people to be treated equal.

Fun Time

How to Tell a Joke

Do you often tell jokes when you speak English? Very few students even try to do it. There are two reasons for this. One is that they have trouble remembering jokes and stories. The second reason is that they don't tell the jokes in a natural, funny way. Now I can give you some advice.

First, enjoy the following joke and try to retell it to your classmates.

A Good Teacher

One day, a teacher was teaching the names of animals to a class of five-year-olds. She held up a picture of a deer, and asked one boy, "Billy, what is this animal?" Little Billy looked at the picture with a disheartened (灰心的) look on his face and answered, "I'm sorry, Mrs. Smith, I don't know." The teacher was not one to give up easily, so she then asked Billy, "Well, Billy, what does your Mommy call your Daddy?" Little Billy's face suddenly brightened up, but then a confused look came over his face, as he asked, "Mrs. Smith, is that really a pig?"

Do you know "punch line"? It is the last line of a joke and is the part that makes the story funny. The boy's surprise answer is what makes people laugh. Maybe you could introduce the "punch line" with more detail and say "Little Billy thought for a moment... still confused. Suddenly a bright look appeared on his face as he asked the teacher..." and then you can give the same punch line "Is that really a pig?"

Be sure to practice jokes before you tell them. Especially practice telling the "punch line" because that's the most important part.

UNIT 2

Market Research

Unit Objectives

After studying this unit, you are able to:
- learn some basic knowledge to do a market research
- understand market analysis reports
- design a questionnaire or an interview
- communicate well with interviewees
- draft a market report

English for Mechanical & Electrical Engineering

Warming-up

Task 1   As a market researcher, Ellen may experience the following situations. Match each situation with its corresponding picture.

☐ Designing a suitable questionnaire.

☐ Asking a potential customer to fill in the questionnaire.

☐ Interviewing an expert on stainless steel (不锈钢).

☐ Presenting a market report in a company's meeting.

Task 2   As a market researcher, Ellen would have to analyze many data or figures. Match the following statements with the corresponding tables or charts.

1. There is a sharp increase in the last five years, that is, from 2015-2020.
2. The past decade witnessed a steady decrease in the labor costs in this field.
3. This year's sales volume has been nearly doubled, namely from ten thousand units to twenty thousand units.
4. There is no substantial change in terms of overall turnout within the past three years.

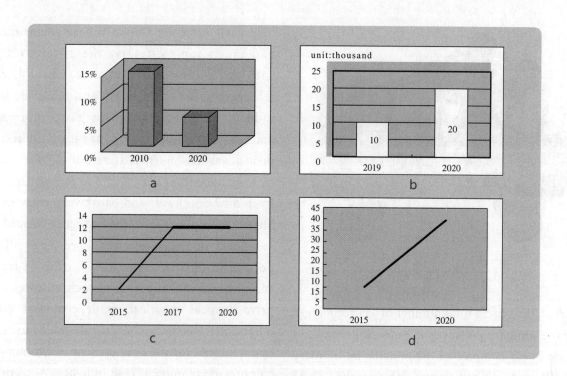

Reading A

Task 1   Before reading the passage, try to answer the questions about market research and market report.

1. Which parts does a market research generally consist of?
2. What makes a market report effective and impressive?

Gas Compressor Market in China — The Past and the Future

As a kind of general machinery, gas compressor is widely used in such industries as petrochemical, steel, metallurgy and automobile. Gas compressor is mainly divided into several types, including centrifugal gas compressor, reciprocating gas compressor, and piston gas compressor.

China's gas compressor manufacturing industry has developed rapidly in recent years. In

Fig 1. Reciprocating gas compressor

Fig 2. Piston air compressor

2007, China had a total number of 355 companies whose annual sales revenue exceeding 5 million *yuan* in gas compressor manufacturing industry and their combined output value reached 56.93 billion *yuan*, amounting to 0.78 of the total output value of the entire machinery industry, the percentage of which was 0.03 percentage points higher than the previous year. Moreover, China had also increased in other indicators like output value of new products, industrial sales output value and export delivery value.

The year 2009 experienced a serious economic depression in the world, but it was too early to say that this industry would also be greatly affected, since Chinese government had taken some effective measures to deal with this problem. At that time, China had a relatively fierce competition and relatively low industry concentration in its gas compressor manufacturing industry. China's gas compressor producers are mainly distributed along the eastern coastal regions, while in its west region (including Southwest and Northwest) there exist huge market potentials to be further developed.

During the Eleventh Five-Year Plan period (2006-2010), the localization of large complete set of equipment in such industries as petrochemical, chemical, textile, coal/electricity/oil and metallurgy would provide huge business opportunities for the development of China's gas compressor manufacturing industry. According to the plan, China would build a great deal of important projects in key industries in the period 2006-2010. Take China's nuclear power industry for instance. More than 20 nuclear power stations will be built in the five years, which will provide not only a broad market for the rapid development of gas compressor manufacturing industry, but also a unique opportunity to accelerate the improvement of the quality of air compressors.

Fig 3. Centrifugal press

Unit 2 Market Research

Task 2 Read the passage and fill in the following chart with figures or numbers given in the passage.

the overall output value of machinery industry in 2007	
the overall output value of gas compressor manufacturing industry in 2007	
annual sales revenue of gas compressor manufacturing industry in 2007	
the percentage of overall output value of gas compressor manufacturing industry in the whole machinery industry in 2007	
the percentage of overall output value of gas compressor manufacturing industry in the whole machinery industry in 2006	

Task 3 Read the passage again and decide whether the following statements are true (T) or false (F).

☐ 1. Gas compressor is not widely applied in various industries in China.
☐ 2. In China, gas compressor manufacturing industry has experienced a rapid increase within these years.
☐ 3. In 2007, the annual sales revenue of gas compressor manufacturing is a little below 5 million *yuan*.
☐ 4. Though the output value of gas compressors was big in 2007, the percentage of the output value of the entire machinery industry decreased that year.
☐ 5. China's main gas compressor producers are equally distributed around China.
☐ 6. The gas compressor manufacturing industry has been well-developed in China, and heated competition occurs in this field.
☐ 7. The Eleventh Five-Year Plan laid great emphasis on the development of many key industries, such as nuclear power, in China.
☐ 8. As many large complete sets of equipment in many industries are being localized, China's gas compressor manufacturers enjoy huge business chances.

Task 4 From the passage, we can see that the author has searched many sources for necessary information. Can you guess what sources may be used in this text? And if you are required to do a desk research (案头调研) about gas compressor, what kind of sources you may use in order to find necessary information?

Listening

Task 1 Listen to the conversation and match the answer with the question.

1. Who is the journalist?
2. Who is the secretary?
3. Who is the person to be interviewed?
4. Who is the person acting as the contact person?

A. Tom Edison
B. Jane Cooper
C. Ellen Johnson
D. Li Jing

Task 2 Ellen Johnson is calling Mr. Li, an expert from China Iron and Steel Association (CISA). But Mr. Li is not in. Listen to the conversation and judge whether the following statements are true (T) or false (F).

( ) 1. Jane is the person Miss Johnson wants to talk with.
( ) 2. Miss Johnson has been to London on a business trip.
( ) 3. Miss Johnson has an appointment with Mr. Li at 9 o'clock the next day.
( ) 4. Miss Johnson is calling to cancel the appointment.
( ) 5. Miss Johnson is very interested in the stainless steel market in India.

Task 3 Ellen Johnson is interviewing Mr. Li. Listen to the conversation and fill in the blanks with what you hear.

Li Jing: Hello, Miss. Johnson, nice to see you.
Ellen: Nice to see you, too, Mr. Li.
Li Jing: Did you have a good trip?
Ellen: Yes, thank you. This trip was really 1 _____ and the weather here is 2 _____.
Li Jing: I'm glad you like the weather here.
Ellen: I really like it. The weather in London now is often 3 _____ and humid.
Li Jing: Yeah, London is famous for that. Well, Miss Johnson, please take a 4 _____.
Ellen: Thank you. Can we start now?
Li Jing: Sure.

Unit 2 Market Research

Ellen: Mr. Li, as we know, China was the world's number one producer of stainless steel last year. Can you 5 _____ our readers of some of the details?

Li Jing: Sure. In 2008, as CISA reported, China took up more than one quarter of the global 6 _____, that is, about 7.2 million tons. And we have 7 _____ Japan as the biggest producer in 2007.

Ellen: But some 8 _____ also showed that in the past half year, steel products exports have 9 _____ sharply. What was the main reason for this decrease?

Li Jing: Well, this is a very difficult problem. But one of the reasons, to be frank, is that the export tax has been 10 _____.

Task 4 Ellen Johnson is continuing her interview with Li Jing. Listen to the conversation and choose the best answer to each question.

1. Which of the following table is suitable to describe the change of the export of crude steel?

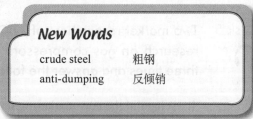

New Words

crude steel   粗钢
anti-dumping   反倾销

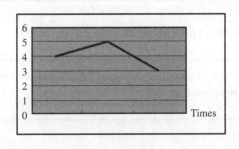

A

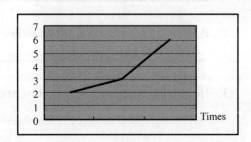

B

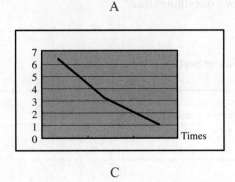

C

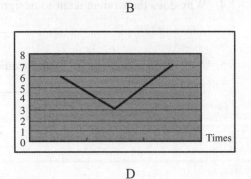

D

2. Which of the following changes is not mentioned in the dialogue?

   A. The export of steel has been affected.

   B. The overall product of steel has been affected.

   C. The import of steel has been affected.

   D. The export of products with advanced technology has increased.

3. What did some European countries do with regard to Chinese steel industry?

   A. They imported more steel from China.

   B. They exported more steel to China.

   C. They complained that Chinese steel industry was dumping.

   D. They wanted to communicate with Chinese steel industry.

4. Which of the following is not an effective way to solve the problem according to the interview?

   A. Dialogs.

   B. Communications.

   C. Negotiations.

   D. Complaints.

Task 5 Two market researchers of ABC Company are discussing about how to do a market research on gas compressors available in the market. Listen to the conversation three times and answer the following questions.

1. What did the new marketing manager ask the two market researchers to do?
   _____

2. How many second-hand information sources are mentioned? What are they?
   _____

3. How many first-hand information sources are mentioned? What are they?
   _____

4. Why does the woman want to design two questionnaires?
   _____

5. According to the woman, when would an expert interview be done?
   _____

Unit 2 Market Research

Speaking

Task 1 Work in pairs. Practice making short phone calls with the words provided according to the example below.

> Example: Li Mei, a journalist from *China Daily*/Mr. Thomson/technology and products/next Monday afternoon

A: Hello, this is Li Mei speaking. Can I speak to Mr. Thomson?

B: Speaking please, Miss Li.

A: Mr. Thomson, I am a journalist from *China Daily*. We'd like to start a new column in our newspaper, which is called "Dialog with Industrialists". I am wondering if you can be our first guest.

B: Sure, it is my pleasure.

A: Thank you so much. Our readers are very much interested in the technology and products of your company. We would appreciate it if you could give a brief introduction of them.

B: OK, no problem.

A: Thank you so much, Mr. Thomson. So when will you be free?

B: Let me have a look at my schedule. I will be free next Monday afternoon. Is that OK?

A: Sure, so I will see you at 2 o'clock next Monday afternoon.

B: See you.

1. Wu Wei, a journalist from *Machinery*/Mr. Wood/steel product/this Friday afternoon
2. Xie Ting, a journalist from *Technology of Mechanical and Electrical Integration*/Mr. Robinson/Siemens' latest products/next Tuesday morning
3. Zhang Bo, a market researcher from TA Industrial Design Company/Mr. Jordan/extrusion mold/May 1

Task 2 Work in pairs. Li Mei is interviewing Mr. Thomson, an analyst from the Marketing Department of ABC Compressor Co., Ltd. Make a conversation according to the instructions below.

1. What is your main product?	A leading manufacturer of compressor/the top native compressor brand
2. What is the annual sales figure of this product?	Approved by ISO9001-9002/Price is favorable.
3. Where are your main markets?	Applies ERP management system.
4. How many of your products are exported every year?	Product: screw compressor, variable speed compressor, piston compressor
5. Does the economic depression affect your company?	Main markets: South America, Eastern Europe
	Total annual sales figures: US$40 million–US$50 million
	Export percentage: 31%–40%

Task 3 Work in pairs. Make a conversation according to the given situation.

Situation:
ABC Compressor Co., Ltd used to be the top gas compressor manufacturer in the local market. But in these two years, this company was not doing well. So the CEO wanted the Marketing Dept. to do a market research to find the causes. So Helen and Mr. Thomson are discussing about how to do this market research.

Related expressions:
1. a sharp decrease in sales volume in these two years
2. major competitors/desk research/field research
3. official websites/company websites/professional publications
4. design questionnaire/data collection and analysis
5. expert interviews/customer interviews

Unit 2 Market Research

Task 4 Work in pairs. Read the sample and make a conversation about how to prepare a presentation with the situation given below.

Sample:

(Helen is talking with Mr. Thomson about the presentation she is preparing.)

Helen: Mr. Thomson, how should I present our sales figures and targets to the investors?

Mr. Thomson: Well, to start with, you should show how quickly the company is growing. A simple column chart showing our total sales over the past five years will be best for that.

Helen: I see. How will I show our sales are well-balanced across all products?

Mr. Thomson: An area chart is the best choice for that.

Helen: OK. Finally, I want my colleagues to have a clear idea of where we're headed. What's the best way to show them our sales targets for the next three years?

Mr. Thomson: I think a line chart would be a perfect one. Since our targets are so distinct, we can make them a 3D chart also.

Helen: Sounds good. Alright, looks like we have settled everything. Thank you.

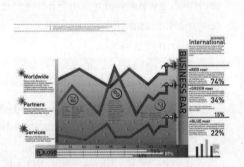
Area chart

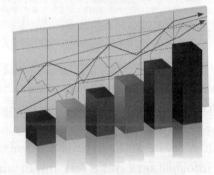
3D bar chart

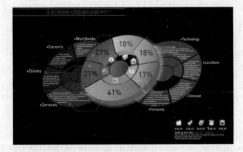
Pie chart

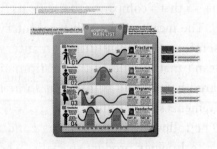
Line chart

Situation:

Mike has done a survey about stainless iron market in China, but he has some problems with data presentation. So he is talking with his manager about how to report data efficiently.

Reading B

Expert Interview—MEMS Switch Applications for Semiconductor Test Market

An interview with an expert is an important element in any industrial market research. From the interview, people can obtain accurate and extensive information. That information can be very important, as a company can get a clear picture of its past performance. And it can also provide necessary results, so that a company can decide its future plans. The means of the industrial interviews are semi-conducted. During an interview, it doesn't matter whether a questionnaire is prepared beforehand or not, but some basic requirements should be met.

The following interview was conducted with a technology product specialist who has an extensive background in semiconductor test engineering, associated software and hardware technologies. This expert has worked with many leading semiconductor providers to understand the needs and challenges of the semiconductor test industry; what's more, he also has made some contributions to successful application and development of innovative software and hardware technologies. Furthermore, this expert has co-founded and provided leadership for companies. These companies are responsible for developing technology products for the semiconductor test market and providing sales, marketing and engineering services.

The interview answers the following questions:

— What are the main applications for MEMS switches in the semiconductor test market?

— What problems can MEMS switches solve in the context of semiconductor testing?

— Who are the target customers?

— Are MEMS switches currently used in any semiconductor test equipment? If so, at which companies and in which equipment and applications?

— What are the main limitations of MEMS switches for semiconductor test equipment applications?

— Can these limitations be overcome and how are they currently being resolved?

— What are the main suppliers of MEMS switches on the market today?

— Which companies are currently targeting the semiconductor test equipment market?

Unit 2 Market Research

Task 1 Read the passage and decide whether the following statements are true (T) or false (F).

☐ 1. Interview plays a very important role in marketing strategies.
☐ 2. From interview, we can get a general understanding of the market or a company's product.
☐ 3. In the majority of industrial interviews, the questions are prepared beforehand.
☐ 4. Generally speaking, there are some basic requirements to follow while an interview is conducted.
☐ 5. The expert mentioned in this passage specializes in test technology and engineering.
☐ 6. The expert owns some companies which develop product technology and provide marketing services.

Task 2 Do you think all the interview questions mentioned in the passage are necessary? If you are the interviewer, what other questions do you think should be listed?

Task 3 Translate the following paragraph in Reading B into Chinese.

> The following interview was conducted with a technology product specialist who has an extensive background in semiconductor test engineering, associated software and hardware technologies. This expert has worked with many leading semiconductor providers to understand the needs and challenges of the semiconductor test industry; what's more, he also has made some contributions to successful application and development of innovative software and hardware technologies.

_____
_____
_____
_____
_____
_____
_____

Writing

Task 1 Read the following mini-report and learn the expressions for depicting graphs and tables.

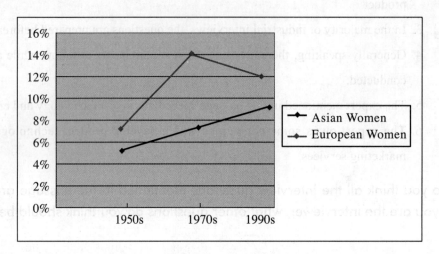

The graph compares the percentage of women smokers in Asia and Europe during the years 1950 to 1990.

It can be clearly seen that from 1950 to 1990 the percentage of women smokers in Asia kept increasing, while in Europe it first rose and then fell.

In the 1950s the percentages of women smokers in Asia and Europe were about the same. From then on, women smokers of both continents began to rise, but the rate of increase in Europe was much bigger than that in Asia. In the 20 years from 1950 to 1970, women smokers of European continent doubled from 7% to 14%, while in Asia, there was only a small rise.

However, since the 1970s, things have changed dramatically. In the following 20 years, the percentage of women smokers in Asia never stopped growing, while in Europe it dropped.

Task 2 Write a mini-report based on the following graph. You can refer to the sample in Task 1.

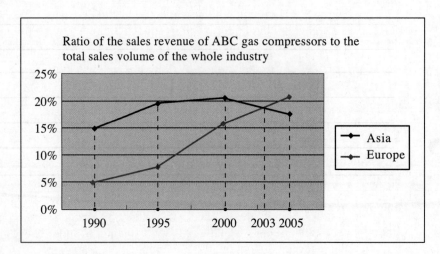

Unit 2　Market Research

Project

Project Guideline

This project aims to go through the procedure of doing a market research on a product. The whole process is divided into four steps. Step One is about desk research. Step Two focuses on designing a questionnaire. Step Three is preparing for an interview. Step Four rests on drafting a mini-report based on the collected information and presenting it to the class.

Please follow the *Task Description* to complete the project.

Task Description

1　Step One
- Organize a group of 6-8 students in your class;
- Decide on which product to be surveyed;
- Discuss with your group members what information you have got.

2　Step Two
- Share the information your group have got with other groups;
- Design a questionnaire based on the information;
- Invite students from another group to fill in the questionnaire.

3　Step Three
- Design some questions for an expert interview;
- Do an interview with some experts or your teacher.

4　Step Four
- Collect and analyze all the data and information;
- Write a mini-report;
- Present your group's market report in the class.

English for Mechanical & Electrical Engineering

Self-evaluation

Rate your progress in this unit.	D	M	P	F*
I can carry out a market research.	☐	☐	☐	☐
I can understand market analysis reports.	☐	☐	☐	☐
I can design a questionnaire or an interview.	☐	☐	☐	☐
I can well communicate with interviewees.	☐	☐	☐	☐
I can analyze the market status.	☐	☐	☐	☐
I can draft a market report.	☐	☐	☐	☐

***Note**: Distinction, Merit, Pass, Fail

New Words and Expressions

Reading A

New Words

accelerate /əkˈsɛləˌret/ v. 促进，加快……的速度
annual /ˈænjuəl/ a. 一年的，全年的
centrifugal /sɛnˈtrɪfjʊgl/ a. 离心的
compressor /kəmˈprɛsɚ/ n. 压缩机，压缩器
delivery /dɪˈlɪvəri/ n. 运送物
distribute /dɪˈstrɪbjut/ v. 分布
exceed /ɪkˈsid/ v. 超过
exist /ɪgˈzɪst/ v. 存在（某种情况）
fierce /fɪrs/ a. 激烈的，极度的
indicator /ˈɪndəˌketɚ/ n. 指标
localization /ˌlokələˈzeʃən/ n. 地方化，本土化
metallurgy /ˈmɛtlˌɝdʒi/ n. 冶金
output /ˈaʊtˌpʊt/ n. 产量
petrochemical /ˌpɛtroˈkɛmɪkl/ n. 石油化学制品
piston /ˈpɪstn/ n. 活塞

previous /ˈpriviəs/ a. （时间或顺序上）先的，前的
reciprocate /rɪˈsɪprəˌket/ v. （指机件）沿直线往复移动
relatively /ˈrɛlətɪvli/ ad. 相对地
revenue /ˈrɛvəˌnu/ n. 收益
textile /ˈtɛkstaɪl/ n. 纺织

Phrases & Expressions

amount to 总计

Technical Terms

centrifugal gas compressor 离心式压缩机
gas compressor 气体压缩机
piston gas compressor 活塞式压缩机
reciprocating gas compressor 往复式压缩机

Reading B

New Words

accurate /ˈækjərət/ a. 准确的，精确的
application /ˌæplɪˈkeʃən/ n. 应用
challenge /ˈtʃælɪndʒ/ n. 挑战性的要求，难题
extensive /ɪkˈstɛnsɪv/ a. 大范围的，大量的
innovative /ˌɪnəˈvetɪv/ a. 创新的，革新的
limitation /ˌlɪməˈteʃən/ n. 局限，限制
overcome /ˌovɚˈkʌm/ v. 战胜，克服（某事物）
resolve /rɪˈzɑlv/ v. 解决（问题、疑问等）
semi-conducted /ˌsɛmɪkənˈdʌktɪd/ a. 半引导的
semiconductor /ˌsɛmɪkənˈdʌktɚ/ n. 半导体

specialist /ˈspɛʃəlɪst/ n. 专家

Phrases & Expressions

be responsible for 对……负责任
make contributions to 对……做出贡献

Technical Terms

MEMS Micro-Electro-Mechanical systems 微机电系统
MEMS switch 微机电系统开关

Vocabulary and Structure

Task 1 Write out the words in Reading A according to their meanings in the right column. The first letters are already given.

d_____	to spread something within in a large area
a_____	to make go faster
e_____	to be greater than; surpass
r_____	money that a business or an organization receives over a period of time
a_____	yearly
p_____	the inherent ability or capacity
f_____	with a lot of energy and sometimes violent
r_____	to solve or settle (problems, doubts, etc.)

Task 2 Fill in each blank with the appropriate form of the word given in the brackets.

1. The committee made a number of (recommend) _____ for improving safety standards.

2. He was (involve) _____ in a heated discussion.

3. First forget those difficult problems. Let's try to deal with a (relative) _____ minor one.

4. The volunteers (contribution) _____ their own time to the project.

5. This new invention will have many (apply) _____ in industry.

6. You must make yourself personally (responsibility) _____ for paying these bills.

7. This is really a (challenge) _____ task for any inexperienced worker.

8. With the development of science and technology, many products, which were imported from other countries, are now being (local) _____.

9. Stress and tiredness often result in a lack of (concentrate) _____.

10. The company decided to launch a sales campaign (target) _____ at the youth market.

Unit 2 Market Research

Task 3 Complete the following sentences with the words or phrases given below. Change the form if necessary.

survey	exceed	conduct	indication	resolve
overcome	amount to	delivery	divide	distribute

1. She _____ that she would never see him again.
2. Dark green leaves are a good _____ of healthy roots.
3. In the 19th century the government _____ land to settlers willing to cultivate it.
4. They _____ the enemy after a long and hard battle.
5. Time lost through illness _____ more than 300 working days.
6. Of the 500 householders _____, 40% had dishwashers.
7. I'm having some flowers _____ for her birthday.
8. Their success _____ all expectations, that is to say, it was greater than anyone expected.
9. The company _____ a survey to find out local reaction to the TV set.
10. The book is _____ into six parts.

Task 4 Make sentences with the same pattern as is shown in the examples.

> *Example 1:* the combined output value of gas compressor manufacturing industry amounted to 0.78 of the total output value of the entire machinery industry/the percentage was 0.03 percentage points higher than the previous year
>
> — The combined output value of gas compressor manufacturing industry amounted to 0.78 of the total output value of the entire machinery industry, the percentage *of which* was 0.03 percentage points higher than the previous year.
>
> *Example 2:* the number of candidates who passed the examination amounts to 81% of the total number of applicators/the percentage was much higher than that of the past year
>
> — The number of candidates who passed the examination amounts to 81% of the total number of applicators, the percentage *of which* was much higher than that of the past year.

1. he has lots of books/most of them are English ones

   _____

2. she has bought many clothes/some of them are second-hand

   _____

3. this project has attracted many investments/some of them came from foreign countries

   _____

4. this underdeveloped country was in great demand of many advanced technologies/most of them were imported from developed ones

   _____

5. this new invention contains many latest technologies/most of them are localized

   _____

> *Example 1:* This interview was conducted with a technology product specialist who has an extensive background in semiconductor test engineering.
>
> — This interview was conducted with a technology product specialist *having* an extensive background in semiconductor test engineering.
>
> *Example 2:* This is a serious accident which involves many people.
>
> — This is a serious accident *involving* many people.

1. This is a very complicated project which needs cooperation and communications among different departments.
   _____

2. Ellen is interviewing an expert who has provided leadership for many companies.
   _____

3. The Forbidden City is a world-famous place of interests which attracts millions of people home and abroad.
   _____

4. The cost has increased rapidly during the next 20 years, which has reached more than 4,000 dollars in 1990.
   _____

5. He is a new college graduate who majors in mechanical engineering.
   _____

Task 5 Translate the following sentences into English using the words or phrases given in the brackets.

1. It is a good little car, but _____ (它有很多明显的不足之处). (limitation)

2. In order to make future plans for product manufacturing, it is necessary to _____ (广泛地调查目标市场). (extensive)

3. _____ (随着技术的本土化), the prices for some automobiles have been reduced dramatically. (localize)

4. The product _____ (在出口市场上甚至会有更大的销售潜力). (potential)

5. The Ferrari Mondial can _____ (从时速0英里加速到60英里) in 6 seconds. (accelerate)

Grammar

Modal Verbs

Task 1 Complete the following sentences by marking the proper model verb with a "√".

1. May/Can you play the piano?
2. Suddenly all the lights went out. We can't/couldn't see anything.
3. You may/can have seen the play already.
4. Could/May you open the window a bit, please?
5. They can/might be away for the weekend, but I'm not sure.
6. Nobody's answering the door. They can/must be out.
7. Will/Shall you tell me where you are coming from?
8. Let's go for a walk, would/shall we?
9. You should/would read this book; it is worth reading.
10. Will/Would you mind giving me a lift if you could?

Task 2 Read the following conversation between a waiter (W) and a customer (C) in a restaurant. Change the underlined words to make the conversation more polite.

> would you like

W: What 1 do you want to order?
C: 2 I want the roast chicken dinner.
W: Anything else?
C: Yes. 3 Bring me a salad.
W: What kind of dressing 4 do you want?
C: 5 Put garlic dressing on my salad.
…
W: Here's your salad, Miss.
C: Thanks. You know, it's a little cold at this table. 6 Let me sit at another table.
W: Of course. There's a nice table in the corner. 7 Sit over there.
C: Thanks, and 8 bring me another glass of water.
W: Of course.

English for Mechanical & Electrical Engineering

Task 3 A husband (H) and his wife (W) are driving to a party but get lost. They are arguing in the car. Fill in the blanks to complete the conversation with the appropriate form of the words given in the brackets.

W: We're lost. And we don't even have a map. You should have brought a map.
H: I didn't think we were going to need one. I must 1 *have made* (make) a wrong turn.
W: Let's use the cell phone to call the Allens and ask them how to get to their house.
H: Where is my phone? I can't find it. I must 2 _____ (leave) it at home.
W: No, you didn't leave it at home. I've got it here in my purse. ... Oh, no. You forgot to recharge the battery. You should 3 _____ (charge) it last night.
H: Why is it my fault? You could 4 _____ (charge) too.
W: Well, we'll just have to look for a pay phone. Do you have any change?
H: I just have dollar bills.
W: You should 5 _____ (bring) some change with you.
H: Again, it's my fault.
W: Watch out! You could 6 _____ (hit) that car!
H: I wasn't going to hit that car. I didn't come anywhere close to it.
W: I don't know why we're going in our car anyway. The Peterson's offered us a ride. We could 7 _____ (go) with them.
H: You should 8 _____ (go) with the Peterson's and I should 9 _____ (stay) home. I could 10 _____ (watch) the football game today instead of listening to you complain.

Task 4 Correct the mistakes in the following letter.

Dear Susan,

Your letter was very interesting, and you are certainly in a difficult situation. I think the first thing you 1 ~~would~~ *should* do is to arrange a meeting with your husband. If I were you, I 2 ~~will~~ _____ tell him exactly how I felt about his mother's behavior. He 3 ~~shall~~ _____ get angry at first, but hopefully he will listen and understand your situation. You could 4 ~~asking~~ _____ him to talk to her and find out why she's always rude to you. If he still doesn't understand your problem, perhaps you 5 ~~should to~~ try to talk to your mother-in-law yourself. And finally, you should stop worrying so much. You should 6 ~~being~~ _____ happy that you have such a good husband.

Yours,
Lily

情态动词

情态动词属于助动词，主要包括can, could, may, might, must, should, ought to, shall, will, would等。情态动词与其他动词连用表示说话人的语气，可表达建议、要求、可能和意愿等。情态动词没有人称和数的变化。

Modals	Examples	Functions
can, could	She can sing. He could ski well before his accident.	能力
can, may, could	You can borrow my dictionary. May I have a menu? Could I use your pen?	许可
will, shall	Will you help me with my work? Shall we go to the theater?	征求意见
will, would	I will do my best to help you. I would like a cup of tea.	意愿
must	You must obey the law.	主观必须
must, could, may, might	They must/could/may/might be at home. The plane can't/couldn't/may not/might not be delayed by the fog. I must/could/may/might have left my key at the school yesterday. He can't/couldn't/may not/might not have heard of the news.	推测
should, ought to	I think you should send her some flowers. You ought to be on time. I should/ought to have taken the job. He shouldn't/ought not to miss the exam.	义务；责备

Comprehensive Exercises

Task 1   Choose the words to complete the sentences below.

1. Her last name is Lemont. She _____.
2. She's wearing a wedding ring. She _____.
3. The house is completely dark. They _____.
4. They said they would come. They _____.
5. I saw them go to the beach. They _____.
6. The teacher sometimes asks questions. He _____.
7. There are a lot of clouds, but it _____.
8. The weather will be fine tomorrow. You _____.
9. She isn't in the music room. She _____.
10. Colin is in a bad mood. He _____.

   a. can't be at home
   c. may not need a coat
   e. might not rain
   g. may be asking questions now
   i. may not feel well

   b. must be swimming now
   d. may be French
   f. could arrive a little late
   h. must be married
   j. couldn't be practicing piano

Task 2   The following is a phone conversation between a woman (W) and a mechanic (M). Choose the correct words to fill in the blanks.

W: This is Cindy Fine. I'm calling about my car.
M: I can't hear you. 1 _____ (Could/Might) you speak louder, please?
W: This is Cindy Fine. Is my car ready yet?
M: We're working on it now. We're almost finished.
W: When 2 _____ (would/can) I pick it up?
M: It will be ready by four o'clock.
W: How much will it cost?
M: $375.
W: I don't have that much money right now. 3 _____ (Can/Should) I pay by credit card?
M: Yes. You 4 _____ (may/might) use any major credit card.

(Later at the mechanic's shop)

M: Your car's ready, ma'am. The engine problem is fixed. But you 5 _____ (may/should) replace your brakes. They're not so good.
W: Do I have to do it right away?
M: No, you don't have to do it immediately, but you should do it within a month or two. If you don't do it soon, you 6 _____ (may/would) have an accident.
W: How much will it cost to replace the brakes?
M: It 7 _____ (will/need) cost about $200.
W: I 8 _____ (will/would) like to make an appointment to take care of the brakes next week. 9 _____ (Can/Will) I bring my car in next Monday?

M: Yes, Monday is fine. You 10 _____ (could/should) bring it in early because we get very busy later in the day.

W: OK. See you Monday morning.

Task 3   Use the words given in the brackets to rewrite the following sentences without changing their meanings.

1. They were able to use a computer. (could)
   _____

2. We should be careful. (ought to)
   _____

3. I realize that it was a terrible experience for you. (must)
   _____

4. It's impossible for Martin to be jogging in this weather. (can't)
   _____

5. It is possible that John did not receive my message. (might)
   _____

6. I suggest we go to the swimming pool. (shall)
   _____

7. I want to have a shower. (would)
   _____

8. I refuse to take any risks. (will)
   _____

9. We don't have to borrow money to buy the house. (need)
   _____

10. The best thing for you to do is not quit your job. (had better)
    _____

Task 4   Correct the mistakes in the following passage.

There was a knock at the door. I opened it and saw a stranger. "Hello, Fred," he cried. "1 ~~Will~~ _____ I come in?" "How do you know my name?" I asked. "We met ten years ago on a train and you gave me your card." "You 2 ~~would~~ _____ be mistaken," I said. "No, I 3 ~~must~~ _____ not," the stranger said. He produced my card: Fred Ames. I could 4 ~~gave~~ _____ it to him ten years ago, but I 5 ~~can't~~ _____ remember it. "I 6 ~~needn't~~ _____ remember you," I said. "We exchanged cards years ago," the stranger said. "You said, 'You 7 ~~would~~ _____ come and stay with us for as long as you like any time you're in London.' I 8 ~~couldn't~~ _____ have waited so long, but I have been so busy and 9 ~~can't~~ _____ spare time. My wife and children are in the car and we wonder if we 10 ~~would~~ _____ stay with you for a month."

Fun Time

Ship It

A software manager, a hardware manager, and a marketing manager are driving to a meeting when a tire blows on the car. They get out of the car and look at the problem.

The software manager says: "I can't do anything about that—it's a hardware problem."

The hardware manager says: "Maybe if we turned the car off and on again, it would fix it itself."

The marketing manager says: "Hey, 75% of it is working—let's ship it."

Borrow Money

Tom: William has asked me for a loan of five pounds. Should I be doing right in lending it to him?

Jack: Certainly.

Tom: And why?

Jack: Because otherwise (否则) he would try to borrow it from me.

UNIT 3

Product Designing

Unit Objectives

After studying this unit, you are able to:
- master some expressions about CAD and gear reducer
- tell the different types of CAD and their functions
- discuss with clients and colleagues about product design
- present designs to the clients
- draft a product design scheme

English for Mechanical & Electrical Engineering

Warming-up

Task 1   As a designer, Steven may experience the following situations. Match each situation with its corresponding picture.

☐ Referring to some books on designing.

☐ Drawing pictures on computer.

☐ Discussing with colleagues for new ideas.

☐ Getting feedback from a client.

Task 2   The following things are what Steven usually does for designing a product. Arrange them in the order of time and explain.

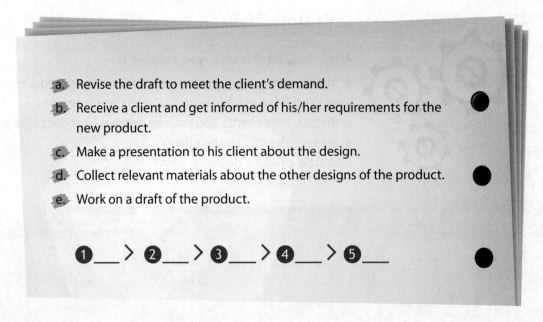

a. Revise the draft to meet the client's demand.
b. Receive a client and get informed of his/her requirements for the new product.
c. Make a presentation to his client about the design.
d. Collect relevant materials about the other designs of the product.
e. Work on a draft of the product.

❶___ > ❷___ > ❸___ > ❹___ > ❺___

Reading A

Task 1 Before reading the passage, try to answer the questions about product designing.

1. What is the complete form of *CAD*? Can you say something about CAD?
2. How can CAD help designers to achieve efficient design?

Computer-aided Design

Computer-aided Design (CAD) is the use of computer technology to aid in the design and particularly the drafting of a part or product. Nowadays, CAD has become an especially important technology, which greatly improves work efficiency compared with hand drafting. It has many benefits such as lowering product development costs and shortening the design cycle. CAD enables designers to lay out and develop work on screen, print it out and save it for future editing, thus saving time on their drawings.

There are several different types of CAD. Each type requires the operator to think differently about how he will use it and design virtual components in a different manner.

There are many producers of the lower-end 2D systems. These provide an approach to the drawing process without the trouble on scale and placement on the drawing sheet that accompanied hand drafting, since these can be adjusted as required during the creation of the final draft. With their help, designers do not need to worry about scale and placement on the drawing sheet.

3D wireframe is basically an extension of 2D drafting. Each line has to be manually inserted into the drawing. The final product has no mass properties associated with it and cannot have features directly added to it.

3D "dumb" solids are created in a way similar to controlling real world objects. Basic three-dimensional geometric forms have solid volumes added or subtracted from them, as if assembling or cutting real world objects. Two-dimensional projected views can easily be generated from the models.

3D parametric solid modeling requires the operator to use what is referred to as "design intent". The objects and features created are adjustable. Any future modifications will be simple, difficult, or nearly impossible, depending on how the original part was created. If a feature was intended to be located from the center of the part, the operator needs to locate it from the center of the model, not, perhaps, from a more convenient edge or any point, as he could when using "dumb" solids. Parametric solids require the operator to consider the consequences of his actions carefully.

In principle, CAD could be applied throughout the design process, but in practice its impact on the early stages has been limited because sketches are widely used at that time. There are some new software programs currently available for these stages. It remains to be seen how effective they will be and how widely they will be implemented.

English for Mechanical & Electrical Engineering

Task 2 Read the passage and match the items on the left with their features on the right.

lower-end 2D systems	an extension of 2D drafting
3D wireframe	shortening the design cycle
3D "dumb" solids	having solid volumes added or subtracted
3D parametric solid modeling	no worry about scale and placement on the drawing sheet

Task 3 Read the passage again and decide whether the following statements are true (T) or false (F).

☐ 1. CAD is the use of computer technology to aid in the design and manufacturing of a product.
☐ 2. With CAD, designers can develop work on screen and save time on their drawings.
☐ 3. In principle, CAD could be applied in the early stages of designing.
☐ 4. Sketches are widely used throughout designing.
☐ 5. CAD is an especially important technology.
☐ 6. 3D "dumb" solids require the operator to consider the consequence of his actions carefully.
☐ 7. In 3D wireframe, each line has to be manually inserted into the drawing.
☐ 8. Features cannot be directly added to the final product of 3D wireframe.
☐ 9. Adding or subtracting solid volumes is like assembling or cutting real world objects.
☐ 10. There are many producers of the 3D parametric solid modeling systems.

Task 4 Discuss with your deskmate about the benefits of application of CAD in industry.

54

Unit 3　Product Designing

🎧 Listening

Task 1　Listen to the conversation and match the people with the correct information.

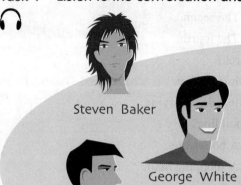

- the general manager of Cascade Elevator Company
- was invited to attend a meeting tomorrow morning
- will be in charge of the design
- visited Cascade Elevator Company
- an experienced designer of elevator system

Task 2　Peter, an intern of Cascade Elevator Company, wants to join the project of designing the elevator system for Carlson Real Estate. Listen to the conversation and fill in the blanks with what you hear.

Peter:　Good morning, Mr. Baker.

Steven:　Morning, Peter. 1 _____?

Peter:　Not much. Oh, Mr. Baker, is that true that you'll have a meeting on the new project half an hour later?

Steven:　Yeah, we'll design the elevator system for the 2 _____ of Carlson Real Estate.

Peter:　Mr. Baker, do you think I can 3 _____ in the project? You know, I've been an intern for several months. And I have a 4 _____ of CAD. Actually I sent an email to you yesterday about it, but there's 5 _____.

Steven:　Sorry, I was too busy to check my mail box. Well, I really need an 6 _____ _____. Maybe you can join us in the first two stages of our design.

Peter:　Oh! That means I will do the 7 _____ and make concept sketches.

Steven:　Also you can 8 _____ the concept color renderings (色彩效果图).

Peter:　Really? Thank you very much, Mr. Baker!

Steven:　That's all right. So, you need to be in the meeting this morning.

Peter:　OK.

Task 3 George White tells Steven Baker his opinions on the concept sketches. Listen to the conversation and choose the best answer to each question.

1. Which stage will the design be in?
   A. The first. B. The second.
   C. The third. D. The fourth.
2. How does George White think of the market research?
   A. It needs improvement. B. It is perfect.
   C. It is quite all right. D. It is dissatisfying.
3. What is the function of the first five floors of the building?
   A. A supermarket. B. A bookstore.
   C. A hotel. D. A shopping mall.
4. What do designers do before they have concept sketches?
   A. They go through the market research carefully.
   B. They meet with the client.
   C. They refer to some books.
   D. They have brainstorming.
5. Why is George White not satisfied with the concept sketches?
   A. Because the elevator system is not efficient.
   B. Because there are several shuttle elevators.
   C. Because they are black-and-white sketches.
   D. Because the shuttle elevators cannot meet his requirements.

New Words
brainstorming *n.* 自由讨论，发表独创性意见
tenancy *n.* 租赁
shuttle elevator 往返直达电梯

Unit 3 Product Designing

Task 4   Steven Baker and Peter are talking about the design. Listen to the conversation and answer the following questions.

1. What is Peter working on?
   _____

2. Why is CAD very important in designing a product?
   _____

New Words

graphic user interface   图形用户界面
user-friendly *a.*   用户容易掌握使用的
refinement *n.*   改进

3. Why is Graphic User Interface necessary?
   _____

4. What should Peter hand in for the part of Graphic User Interface?
   _____

5. Are the graphic design refinements made only in the second stage of design?
   _____

Task 5   Steven Baker presents the design to George White. Listen to the short passage and take notes of the information you hear.

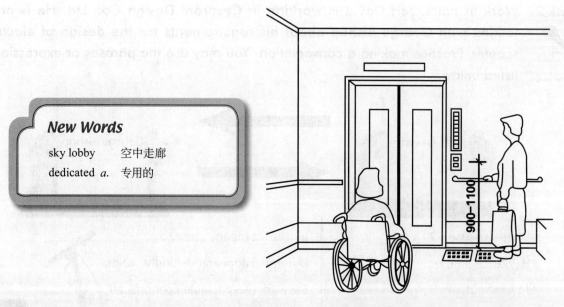

New Words

sky lobby   空中走廊
dedicated *a.* 专用的

   Steven and his colleagues not only designed six lower-speed elevators in the shopping mall area, but also planned two 1 _____, 2 _____ shuttle elevators from the main lobby to the sky lobby. Passengers on these elevators can have dedicated 3 _____ 4 _____ their tenancy types. The "5 _____ 6 _____ 7 _____" systems are used in the elevators to enable people to get to their destinations most efficiently.

57

English for Mechanical & Electrical Engineering

Speaking

Task 1 Work in pairs. Practice making short phone calls with the words provided according to the example below.

> Example: Steven Baker/Cascade Elevator Company/George White/elevator/market research
>
> **A:** Hello, this is Steven Baker of Cascade Elevator Company. Speaking, please.
> **B:** Good morning, Mr. Baker. This is George White. I'm calling to ask how you proceed with the design of our elevator.
> **A:** Mr. White, we've finished the concept sketches. And we'll submit the sketches to you today.
> **B:** OK. I can't wait to see them.

1. Jeff David/Giant Bicycles Co., Ltd./George Bishop/electric scooter/concept color renderings
2. Fang Dawei/Galanz Electrical Equipment Co., Ltd./Susan Bird/gas compressor/concept sketches
3. Joan Robinson/TA Industrial Design Company/Pat Jordan/extrusion mold/2D graphic design

Task 2 Work in pairs. Jeff David is working in Cesaroni Design Co., Ltd. He is now talking with George Bishop about his requirements for the design of electric scooter. Practice making a conversation. You may use the phrases or expressions listed below.

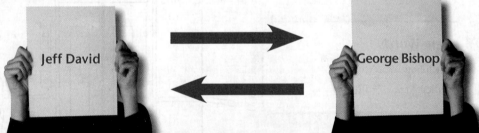

Where to use…?	In the company area.
I'd like to know…	Dynamic appearance/bright color.
How about…?	Easy to use/portable/economical/safe enough.
What about…?	Half a year.
Are there any other requirements?	A timely report.

58

Unit 3 Product Designing

Task 3 Work in pairs. George Bishop is talking with Jeff David about how he felt about the design. He is generally satisfied with it but thinks that some parts need refinement. Make a conversation according to the instructions below.

 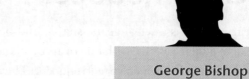

Jeff David

Greet.

Ask about George Bishop's opinions on the design.

Tell him the electric scooter weighs 25 pounds with the battery, and it can be easily folded up.

Promise to have refinement to meet his requirements.

George Bishop

Greet.

Tell Jeff he is satisfied with the draft, but the electric scooter seems not portable enough.

Insist the electric scooter should be lighter.

Look forward to the new electric scooter design.

Task 4 Work in pairs. Jeff David is now presenting the prototype of electric scooter to George Bishop. Make a conversation with your partner according to the information given below.

George Bishop

- Ask about the appearance.
- Want to know about the speed.
- Enquire about weight.
- Wonder whether there is any other information.
- Show your satisfaction.

Jeff David

- Bright orange color; 34 inches long, 7 inches wide, and 35 inches from deck to handlebars; tough, aluminum body
- 15 miles per hour at most
- Light enough to be carried into the office; 19.5 pounds in weight; easy to fold up
- Low deck height; easy kick scooting; completely redesigned motor controller; improved performance

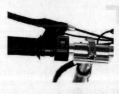 

Reading B

Gear Reducer—A Great Design

A gear reducer is a set of gears, shafts, and bearings that are enclosed in a housing. They are arranged like a solar system, with one or more planet gears orbiting around a sun gear.

Gearboxes are known as speed reducers. They convert input speed into a lower output speed while correspondingly creating higher torque. In other words, gear-boxes reduce RPM, turning it into power for use in low-RPM high-torque applications.

The gear reducer arrangement is a great engineering design that offers many advantages over traditional gearbox arrangements. One advantage is its combination of both compactness and outstanding power transmission efficiencies. The efficiency loss in a gear reducer arrangement is only 3% per stage. This type of efficiency ensures that a high proportion of the energy being inputted into the gearbox is multiplied and transmitted into torque, rather than being wasted on mechanical losses inside the gearbox.

Another advantage of the gear reducer arrangement is load distribution. Because the load being transmitted is shared between multiple planets, torque capability is greatly increased. The more planets in the system, the greater the load ability.

The gear reducer arrangement also creates greater stability (it's a balanced system) and increased rotational stiffness. However, the planetary arrangement has disadvantages. The design is complex and the use of it is too much trouble, for example. In the drawing below, the Fixed Axis Gear System is the traditional arrangement where a pinion is driving one large gear on a parallel shaft. In the Planetary Gear System arrangement, one or more gears (planet gears) surround the pinion (sun gear). Obviously, the Planetary Gear System works better than the fixed Axis Gear System due to its increased number of gear contacts.

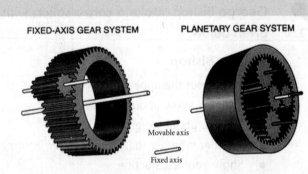

Task 1 Read the passage and match each paragraph with its corresponding main idea.

Paragraph 1 •   • one advantage of the gear reducer
Paragraph 2 •   • what is a gear reducer
Paragraph 3 •   • a comparison of the two arrangement systems
Paragraph 4 •   • the function of gearboxes
Paragraph 5 •   • load distribution of the gear reducer

Task 2 Match the English expressions with their Chinese meanings.

1. planet gear
2. sun gear
3. speed reducer
4. high-torque
5. load distribution
6. torque density
7. rotational stiffness
8. parallel shaft

a. 恒星齿轮
b. 高扭力
c. 行星齿轮
d. 减速器
e. 转动刚度
f. 负荷分配
g. 平行轴
h. 转矩密度

Task 3 Translate the following paragraph in Reading B into Chinese.

> The gear reducer arrangement is a great engineering design that offers many advantages over traditional gearbox arrangements. One advantage is its combination of both compactness and outstanding power transmission efficiencies. The efficiency loss in a gear reducer arrangement is only 3% per stage. This type of efficiency ensures that a high proportion of the energy being input into the gearbox is multiplied and transmitted into torque, rather than being wasted on mechanical losses inside the gearbox.

Writing

Task 1   Suppose you are Peter. When you learn there will be a new project of designing the elevator system for a high building, you write an email to Steven Baker based on the outline below:

1. express your strong desire to participate;
2. tell Mr. Baker you have joined in two projects and are experienced;
3. show your good command of CAD.

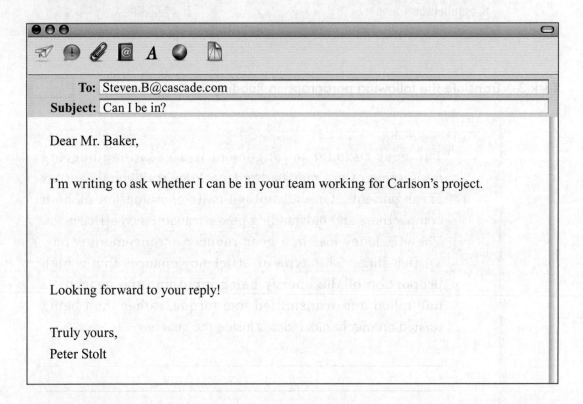

To: Steven.B@cascade.com
Subject: Can I be in?

Dear Mr. Baker,

I'm writing to ask whether I can be in your team working for Carlson's project.

Looking forward to your reply!

Truly yours,
Peter Stolt

Task 2   Joan Robinson, a designer of TA Industrial Design Company, wrote a design plan on a gear system. The following paragraph about the gear system's features is subtracted from the mini-report. Fill in the blanks.

　　Here are the 3D geometric forms of the gear system. One advantage of the arrangement is that it combines _____ (紧密性) with _____ _____ (出色的动力传输效能). _____ (每一阶段只损失3%的效能) ensures that a high proportion of the energy is multiplied and transmitted into torque. Another advantage is load distribution. Torque capability is greatly increased since _____ (传输的负荷由多个行星齿轮共享). In addition, this kind of arrangement has _____ (更大的稳定性) and increased rotational stiffness. To sum up, the Planetary Gear System will create more work efficiency.

Project

Project Guidelines

This project aims to go through the procedure of designing an electric bike. You may have other options for the project. The whole process is divided into four steps. Step One is to know the client's requirements for the product. Step Two is to have a brainstorming meeting in the design group and present the draft to the client. Step Three is to seek the client's feedback and make corresponding refinement. Step Four is about presenting the final design in the class.

Please follow the *Task Description* to complete the project.

Task Description

 Step One

- Set up a group of 4-6 students in your class; select 1-2 students in the group to play the role of client.
- Discuss with your client about his/her requirements for the product.
- Note down the client's requirements. (A thorough understanding of the client's requirements may contribute to your group's victory in the project.)

 Step Two

- Log onto the Internet or go to the library to collect information of the like products available in the market.
- Have brainstorming; put forward suggestions about how to design the product.
- Present the draft of the design to the client.

 Step Three

- Ask about the client's opinion of the draft.
- Have discussion in the group about how to improve the draft.
- Make corresponding refinement.

 Step Four

- Present the last design to the client.
- The client decides which design can make him/her most satisfied and explains why.

Self-evaluation

Rate your progress in this unit.	D	M	P	F*
I can describe the different types of CAD and their functions.	☐	☐	☐	☐
I can communicate with a client about the design requirements of a product.	☐	☐	☐	☐
I can discuss the design of a product with my colleagues.	☐	☐	☐	☐
I can read a product design scheme.	☐	☐	☐	☐
I can make a presentation of the design to the client.	☐	☐	☐	☐
I can draft a product design scheme or a mini-report.	☐	☐	☐	☐

***Note**: Distinction, Merit, Pass, Fail

New Words and Expressions

Reading A

New Words

accompany /ə'kʌmpənɪ/ v. 伴随，和……一起发生
adjustable /ə'dʒʌstəbl/ a. 可调整的，可调校的
assemble /ə'sɛmbl/ v. 装配，组合
associate /ə'soʃɪˌet/ v. 使发生联系，联合
component /kəm'ponənt/ n. 组成部分，元件
draft /dræft/ n. 草图，设计图
edge /ɛdʒ/ n. 边，棱，边缘
edit /'ɛdɪt/ v. 编辑
enable /ɪn'ebl/ v. 使能够，使可能
geometric /ˌdʒɪə'mɛtrɪk/ a. 几何的，几何学的
implement /'ɪmpləmɛnt/ v. 完成，实现
insert /ɪn'sɜt/ vt. 插入，嵌入
low-end /'loɛnd/ a. 低端的，低档的
manually /'mænjʊəlɪ/ ad. 手动
modification /ˌmɑdəfə'keʃən/ n. 更改，改变，修改
parametric /ˌpærə'mɛtrɪk/ a. 参数的，参量的
placement /'plesmənt/ n. 布局
scale /skel/ n. 规模，范围

sketch /skɛtʃ/ n. 草图，素描 v. 描绘略图
solid /'sɑlɪd/ n. 实体
subtract /səb'trækt/ v. 减去，扣除
virtual /'vɜtʃuəl/ a. 虚拟的
wireframe /'waɪrfrem/ n. 线框，框架

Phrases & Expressions

be compared with 与……相比较
in principle 原则上，基本上
lay out 设计，布局
refer to... as 把……称作，把……当作

Technical Terms

3D dumb solid 三维块体
3D parametric solid modeling 三维参数化实体建模
mass properties 质量属性
solid volume 实体积
two-dimensional projected views 二维投影视图

Reading B

New Words

bearing /'bɛrɪŋ/ n. 轴承
compactness /kəm'pæktnɪs/ n. 密实度，紧密
convert /kən'vɜt/ v. 转变，转化
density /'dɛnsətɪ/ n. 密度，稠密
distribution /ˌdɪstrə'bjuʃən/ n. 分配，分发
enclose /ɪn'kloz/ v. 装入，圈起，封闭
gear /gɪr/ n. 齿轮，传动装置
gearbox /'gɪrˌbɑks/ n. 变速箱，变速器
housing /'haʊzɪŋ/ n. 外壳，外罩
multiply /'mʌltəˌplaɪ/ v. 增加，乘
orbit /'ɔrbɪt/ v. 环绕轨道运行
outstanding /ˌaʊt'stændɪŋ/ a. 出色的

pinion /'pɪnjən/ n. 小齿轮
planet /'plænɪt/ n. 行星
RPM (abbr.) revolutions per minute 转数/分
shaft /ʃæft/ n. 轴
solar /'solɚ/ a. 太阳的
stiffness /'stɪfnɪs/ n. 坚硬，硬度
torque /tɔrk/ n. (机器的) 扭转力，扭矩，转矩
transmit /træns'mɪt/ v. 传动，传输

Phrases & Expressions

a high proportion of 很大一部分
due to 由于，因为
in other words 换句话说

Vocabulary and Structure

Task 1 Write out the words in Reading A or Reading B according to their meanings in the right column. The first letters are already given.

s_____ to take a part or amount from a larger number or amount

e_____ to make something possible

i_____ to carry out or put into practice

a_____ to put together

a_____ to make a connection between one thing or person and another

m_____ a small change made in something such as a design, plan, or system

c_____ to (cause something or someone to) change in form, character or opinion

i_____ to put something inside or into something else

Task 2 Fill in each blank with the appropriate form of the word given in the brackets.

1. Tom received a (compactness) _____ package, which was delivered by his mother abroad.

2. He (adjustable) _____ himself very quickly to the heat of the country.

3. He got a job as (edit) _____ of a trade journal.

4. The industrial revolution (modification) _____ the whole structure of English society.

5. I cannot play the piano like I used to—my fingers have gone (stiffness) _____ from lack of practice.

6. The new secretary is a quick, (efficiency) _____ worker, and the boss is quite satisfied with her.

7. (density) _____ fog is covering roads and visibility is very poor.

8. Citizens may have free (accessibility) _____ to the library.

9. The World Cup Final is being (transmission) _____ live to over 50 countries.

10. The meeting with the German (distribution) _____ has been advanced from 11:00 to 9:30.

Unit 3 Product Designing

Task 3 Complete the following sentences with the words or phrases given below. Change the form if necessary.

| lay out | apply | convert | associate | implement |
| in a way | insert | subtract | refer to... as | depend on |

1. The cities of Wuchang, Hankou and Hanyang are often _____ Wuhan.
2. The government is _____ a new policy to help the unemployed.
3. You should dress _____ that benefits a woman of your position.
4. _____ to gas central heating will save you a lot of money.
5. Since you have graduated, you should find a job and end your _____ your parents.
6. Please follow the instruction on Page 1 to _____ a diskette into one of your computer's disk drives.
7. The architect spent three months _____ the interior of the building.
8. I was told by my teacher that the rule could not be _____ to every case.
9. We want further cooperation as our long _____ with your company has brought great benefits.
10. Six _____ from nine gives three.

Task 4 Make sentences with the same pattern as is shown in the examples.

> *Example 1:* this type of efficiency ensures that a high proportion of the energy being input into the gearbox is multiplied and transmitted into torque/waste on mechanical losses inside the gearbox
>
> — This type of efficiency ensures that a high proportion of the energy being input into the gearbox is multiplied and transmitted into torque, *rather than* being wasted on mechanical losses inside the gearbox.
>
> *Example 2:* she is a career woman/a housewife
>
> — She is a career woman *rather than* a housewife.

1. we will have the meeting in the classroom/in the auditorium
   _____

2. he should be rewarded/punish
   _____

3. I decided to write/telephone
   _____

4. we should help him/he should help us
   _____

5. she is charming/beautiful
   _____

> *Example 1:* many planets in the system/the great load ability
>
> — <u>*The more*</u> planets in the system, <u>*the greater*</u> the load ability.
>
> *Example 2:* he is busy/he feels happy
>
> — <u>*The busier*</u> he is, <u>*the happier*</u> he feels.

1. you work hard/you will make great progress
   _____

2. you run fast/it will be good
   _____

3. he worked hard/he got much
   _____

4. he worried little/he worked well
   _____

5. you are careful/you will make few mistakes
   _____

Task 5 Translate the following sentences into English using the words or phrases given in the brackets.

1. _____ (由于投资少), our industrial output has remained stagnant. (due to)

2. The Planetary Gear System arrangement is _____ (被称作一项伟大的工程设计). (refer to as)

3. The solar cell can _____ (把阳光的能量转化为电能). (convert)

4. The designer used a lot of sketches on the early stages _____ (尽管他知道整个设计过程都可用到CAD). (apply)

5. I gave him full directions to _____ (好让他能找到那所房子). (enable)

Grammar

Tenses

Task 1 Cross out the incorrect tenses of the verbs in the following sentences.

1. I read/am reading a very interesting book.
2. My wife prefers/is preferring coffee for breakfast.
3. What have you done/were you doing when the accident occurred?
4. Have you ever visited/Do you ever visit Italy?
5. It rained/has been raining all week. I hope it stops by Saturday.
6. When I got there, some people arrived/had arrived before me.
7. Look at the clouds. It is going to rain/will rain soon.
8. When you arrive/will arrive in Shenzhen, I will show you around this city.
9. By the time you get back, I have painted/will have painted the house.
10. They promised they would have finished/will have finished the project by the end of this week.

Task 2 Fill in the blanks with the simple present or present continuous tense of the verbs.

Mark: Hi Julie. Is Bob there?
Julie: He can't come to the phone. He's planting some flowers in the garden.
Mark: But it 1 *is raining* (rain)!
Julie: Bob always 2 _____ (plant) flowers on rainy days. It's perfect weather for planting.
Mark: I suppose you're right. I want to buy a used car. I'm sure Bob 3 _____ (know) something about cars.
Julie: Well, Mark. Bob 4 _____ (come) into the house now. Hey, Bob. Mark's on the phone. He wants to buy a used car.
Bob: Hey, Mark. How's it going? Is it true that you 5 _____ (look for) a used car?
Mark: Sure is. I 6 _____ (need) a car to get a job. Any suggestions?
Bob: My brother Bill 7 _____ (sell) his car. It's a 2006 Honda Civic.
Mark: Sounds interesting. I suppose he 8 _____ (buy) a new car?
Bob: Yeah.
Mark: Do you know how much he 9 _____ (ask for) his Civic?
Bob: No, but I can ask him this evening. Julie and I 10 _____ (go out) to dinner with him.
Mark: Thank you very much.
Bob: You're welcome. Talk to you later.

Task 3 Fill in the blanks with the simple past or present perfect tense of the verbs.

A: I haven't seen you for a long time. Where have you been?
B: I 1 *have just come* (just, come) back from Canada.
A: Oh really? What 2 _____ (you, do) in Canada?
B: I took a nature tour.
A: Wow! 3 _____ (you, see) many wild animals there?
B: Of course. I 4 _____ (watch) bears, wolves and whales in the wild. That was so interesting. 5 _____ (you, ever, spend) a holiday in Canada?
A: Yes, I 6 _____ (travel) around Canada twice so far.
B: When 7 _____ (you, go) there?
A: The first time I 8 _____ (go) there was in 1997 and the second time in 2004.
B: 9 _____ (you, enjoy) it?
A: I absolutely 10 _____ (love) it, especially the west coast.

Task 4 Fill in the blanks with the present perfect or present perfect continuous tense of the verbs.

Robin: I think the waiter has forgotten us. We 1 *have been waiting* (wait) here for over half an hour and nobody 2 _____ (take) our order yet.
Michele: I think you're right. He 3 _____ (walk) by us at least twenty times. He probably thinks we 4 _____ (order, already).
Robin: Look at that couple over there, they 5 _____ (be, only) here for five or ten minutes and they already have their food.
Michele: He must realize we 6 _____ (order, not) yet! We 7 _____ (sit) here for over half an hour staring at him.
Robin: I don't know if he 8 _____ (notice, even) us. He 9 _____ (run) from table to table taking orders and serving food.
Michele: That's true, and he 10 _____ (look, not) in our direction once.

时态

一般现在时 （simple present） 与现在进行时 （present continuous）	一般现在时表示一般行为或不断重复发生的行为：I watch television every day. 现在进行时表示现在正在发生的行为：I'm watching television now.
	表示感觉、情感、存在等动词，如：see, feel, hear, smell, taste；like, love, hate, prefer；exist, appear, seem 等，一般不用现在进行时，而用一般现在时。
	某些转移性动词或瞬间动词，如：come, go, arrive, leave, start, return, meet, get 等，可以用现在进行时表示将来：The train is arriving soon.
一般过去时 （simple past） 与现在完成时 （present perfect）	一般过去时表示过去某一时间所发生的动作或存在的状态，与现在无联系，常与表示过去的时间词连用：I had supper an hour ago. 现在完成时的动作发生在过去，但对现在有影响或产生结果，常与 today, just, recently, since, for, ever 等词连用： I have just had supper.（刚吃完晚饭，现在不饿。）
现在完成时与现在完成进行时 （present perfect continuous）	现在完成时表示已完成的动作。 现在完成进行时表示动作从过去某一时间开始一直延续或断断续续重复到现在。动作是否继续进行下去，由上下文而定： What have you been doing all this time? You've been saying that for five years.
will 与 be going to	be going to 表示近期、眼下就要发生的事情，will 表示的将来时间则较远一些： He is going to write a letter tonight. He will write a book one day.
	be going to 表示根据主观判断将来肯定发生的事情，will 表示客观上将来势必发生的事情： He is seriously ill and is going to die. He will be 20 next month.
	be going to 含有"计划，准备"的意思，而 will 则没有这个意思： She is going to sell her car. He will come if he has time.

Comprehensive Exercises

Task 1 Fill in the blanks with the simple past or past continuous tense of the verbs.

A: What 1 _____ (you, do) yesterday at 8 pm?
B: I 2 _____ (sit) in the pub with Sam. Why?
A: I 3 _____ (drive) to the sports center at that time to play tennis with a few friends. As we had only three players, I 4 _____ (try) to ring you to ask if you would like to come as well. But I 5 _____ (reach, not) you at home.
B: Why 6 _____ (you, not, ring) my mobile?
A: I actually 7 _____ (want) to ring your mobile, but by accident I 8 _____ (dial) William's number. He 9 _____ (do, not) anything special at that moment and really 10 _____ (like) the idea of playing tennis with us.

Task 2 Rewrite the underlined parts using "will" or the correct form of "be going to".

1. My radio is broken. <u>You (fix) it for me?</u>
_____

2. I have a two-week vacation in August. <u>I (fly) to London.</u>
_____

3. Marie is sick. <u>She (not go) with us for the concert.</u>
_____

4. My supervisor is a really nice guy. He promised me a big bonus this year. <u>He promised me that he (give) me $2,500 if we increase sales by 10%.</u>
_____

5. You spilled the milk. <u>I (clean) it up.</u>
_____

6. **A:** Why are you holding a piece of paper?
   B: <u>I (write) a letter to my friends back home in Texas.</u>
_____

7. **A:** I'm about to fall asleep. I need to wake up!
   B: <u>I (get) you a cup of coffee. That will wake you up.</u>
_____

8. **A:** I can't hear the television!
   B: <u>I (turn) it up so you can hear it.</u>
_____

9. We are so excited about our trip next month to France. <u>We (visit) Paris, Nice and Grenoble.</u>
_____

10. **A:** Excuse me, I need to talk to someone about our hotel room. I am afraid it is simply too small for four people.
    B: <u>That man at the service counter (help) you.</u>
_____

Task 3 Complete the text with the past perfect or past perfect continuous tense of the verbs.

I'm sorry I left without you last night, but I told you to meet me early because the show started at 8:00. I 1 _____ (try) to get tickets for that play for months, and I didn't want to miss it. By the time I finally left the coffee shop where we were supposed to meet, I 2 _____ (have) five cups of coffee and I 3 _____ (wait) over an hour. I had to leave because I 4 _____ (arrange) to meet Kathy in front of the theater.

When I arrived at the theater, Kathy 5 _____ (pick, already) up the tickets and she was waiting for us near the entrance. She was really angry because she 6 _____ (wait) for more than half an hour. She said she 7 _____ (give, almost) up and 8 _____ (go) into the theater without us.

Kathy told me you 9 _____ (be) late several times in the past and that she would not make plans with you again in the future. She mentioned that she 10 _____ (miss) several movies because of your late arrival. I think you owe her an apology. And in the future, I suggest you be on time!

Task 4 Correct mistakes concerning verb tenses in the following application letter.

Dear Sirs,

I would like to apply for the job of sales manager that I 1 ~~see~~ __saw__ advertised in the local paper.

I am 30 years old. I was born in Canada but my family 2 ~~had moved~~ __moved__ to New York when I was 10 and I 3 ~~am living~~ __have lived__ here ever since. I 4 ~~have left college for eight years~~ __left college eight years ago__ and since then I 5 ~~am having~~ __have had__ several jobs in sales. For the past three years I 6 ~~am working~~ __have been working__ in a supermarket. The manager 7 ~~has been saying~~ __has said__ that he is willing to give me a reference.

I 8 ~~am speaking~~ __speak__ French and English fluently and I 9 ~~have learnt~~ __have been learning__ German since I 10 ~~had left~~ __left__ school, so I speak some German too.

I hope you will consider my application.

Yours sincerely,
James Brit

Fun Time

The Funny English Language

English seems to be a "stupid" language. Look at the following sentences.

There is no egg in the eggplant. (茄子)
No ham in the hamburger (汉堡).
And neither pine (松木) nor apple in the pineapple (菠萝).
English muffins (英式松饼) were not invented in England.
French fries (土豆条) were not invented in France.

No wonder the English language is so very difficult to learn. you may wonder how you can manage to communicate at all!

Please enjoy the following jokes about the English language. Do you know why they made mistakes?

1

Three students who study English as a foreign language, are walking down the road to their remedial listening comprehension workshop.

"It's windy," says the first.

"No, it isn't. It's Thursday," says the second.

"Me too," says the third. "Forget the listening; let's go for a drink!"

2

A student, who is studying English as a foreign language, was confused when he saw the words "open here" on a box of laundry soap, so he asked the clerk, "Can't I wait until I get home to open it?"

3

My student who did not speak much English wanted to impress me one day. She had to walk past me while I was talking to someone. She said, "Excuse me, can I pass away?"

UNIT 4

Production

Unit Objectives

After studying this unit, you are able to:
- describe the functions and applications of CAM
- understand the engineer's instructions
- discuss with a technician about product making
- introduce the production status to guests
- write a notice and the DPR

English for Mechanical & Electrical Engineering

⚙ Warming-up

Task 1   Work in pairs. Discuss which are electromechanical products in the following pictures.

electrical bicycle   boiler   roller coaster   crane

mobile phone   electric shaver   revolving door   bicycle

Task 2   The following things are what Bob usually does in manufacturing a product. Arrange them in the order of time and explain.

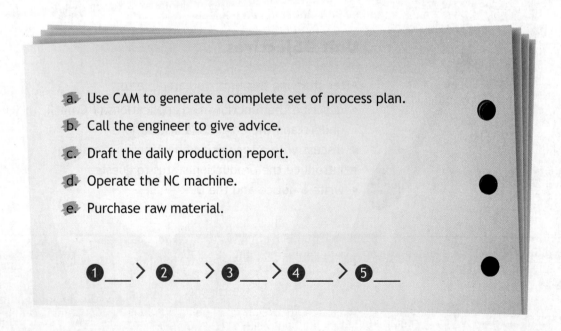

a. Use CAM to generate a complete set of process plan.
b. Call the engineer to give advice.
c. Draft the daily production report.
d. Operate the NC machine.
e. Purchase raw material.

❶___ > ❷___ > ❸___ > ❹___ > ❺___

Reading A

Task 1 Before reading the passage, try to answer the questions about CAM.

1. What's the complete form of *CAM*? Can you say something about CAM?
2. Robots are not easily separated from CAM. What tasks can they handle in industrial automation?

Computer-aided Manufacturing

Computer-aided Manufacturing (CAM) can be defined as the use of computer systems to plan, manage, and control the operations of a manufacturing plant through either direct or indirect computer interface with the plant's production resources.

CAM functions center around four main areas: numerical control, process planning, robotics, and factory management.

Numerical Control (NC)

Numerical control uses coded information to control machine tool movements. In modern CNC (computer numerical control) systems, end-to-end component design is highly automated using CAM programs. The programs produce a computer file that is interpreted to extract the commands needed to operate a particular machine, and then loaded into the CNC machines for production.

Process Planning

Process planning is involved with the detailed sequence of production steps from start to finish. Essentially, the process plan describes the state of the workplace at each workstation. The use of computers as an aid to process planning is comparatively recent and has led to a rebirth of what is known as group technology (GT). Group technology is based on organizing all similar parts into families to allow standardization of manufacturing steps.

Currently a process planning system is under development and it is able to produce process plans directly from the geometric model database with almost no human assistance. In this system, the process planner would review the impact from the design engineer via communication and then enter this input into the CAM system which would generate a complete set of process plans automatically.

Robotics

Many advances are being made to integrate robotics into CAM. One of these efforts is the US Air Force Integrated Computer-aided Manufacturing (ICAM) project, of which the

goal is to organize every step of manufacturing around computer automation. As part of this program, a robot is used to drill sheet metal aircraft parts. The robot drills a set of holes to 0.005 in tolerance, and machines the perimeter of any one of 250 types of parts. Production rates are four times faster than conventional manual manufacturing.

Factory Management

This portion of CAM ties together the other areas to coordinate operations of an entire factory. The management system relies heavily on group technology with its families of similar parts. Computers also perform various management tasks such as inventory control and material requirements planning (MRP) systems.

Task 2 Read the passage and find the corresponding English equivalents for the following Chinese.

Chinese	English
计算机辅助制造	
计算机接口	
几何模型	
成组技术	
金属薄片	
生产率	
数据库	
人工生产	
存货管制（或盘仓）	

Task 3 Read the passage again and match the items on the left with their functions on the right.

1. Numerical Control	a. generate a complete set of process plans automatically
	b. integrate robotics into CAM
2. Process Planning	c. tie together the other areas to coordinate operations of an entire factory
3. Robotics	d. organize all similar parts into families
	e. is involved with the detailed sequence of production steps from start to finish
4. Factory Management	f. use coded information to control machine tool movements
	g. produce process plans directly

Task 4 Discuss in groups. How is CAM applied in the operations of a manufacturing plant? What do you think the benefits of CAM are?

Unit 4   Production

Listening

Task 1 Listen to the conversation and match the people with the correct information.

Jessica

- can't keep appointment on Tuesday
- a student practicing in workshop
- has some questions about CAM system
- will go to Head Office
- will have lunch at the Atlas

Andrew Johnson

Task 2 Listen to the conversation and tick the items mentioned in the conversation.

☐ Ensuring the appointment time.
☐ Changing the appointment time.
☐ Reason of changing the time.
☐ Not available the whole week.
☐ Free in Friday afternoon.
☐ Another appointment time.
☐ The appointment place.

Task 3 Mr. Clive from an American company is visiting Blue-sky Mold, accompanied by Andrew Johnson, the Production Manager. Listen to the conversation and fill in the blanks with what you hear.

New Words
subcontractor n.   次承包商, 转包人
laboratory n.   实验室

Andrew Johnson: Now this is the 1 _____ line that produces the machine tools, Mr. Clive.
Mr. Clive: I've been very impressed by what I've seen. I know the factory is 2 _____ at full capacity.
Andrew Johnson: Yes, we've received plenty of orders, both for 3 _____ needs and for export. As you know, we 4 _____ machine tools of all types and sizes.
Mr. Clive: Is any work done by subcontractors?
Andrew Johnson: No, we are fully self-sufficient. We have laboratories, quality control

79

English for Mechanical & Electrical Engineering

                              department and 5 _____ department all here.

Mr. Clive: Have you been producing this new model for a long time?

Andrew Johnson: Yes, we've introduced new technology and started a new model last year. Our designers always keep up with the 6 _____ of technology.

Mr. Clive: Is the 7 _____ of engineers big?

Andrew Johnson: Totally we have about 200 engineers.

Mr. Clive: How do you ensure quality control?

Andrew Johnson: Well, it's done by the Quality Control Department. Our quality control engineers ensure that all equipment manufactured is thoroughly inspected and 8 _____ in full the requirements of the orders technically.

Mr. Clive: Do you also 9 _____ the packing?

Andrew Johnson: Yes, but we've recently started to use packing companies too. Well, is there anything else you'd like to see?

Mr. Clive: No, thanks.

Andrew Johnson: OK. Let's go to my office for a cup of 10 _____ .

Task 4

Listen to the conversation in Andrew Johnson's office and judge whether the following statements are true (T) or false (F).

( ) 1. Mr. Clive would not like black tea.

( ) 2. Andrew Johnson has worked out a 10-day schedule for Mr. Clive.

( ) 3. Mr. Clive thinks Peking Opera is a little different from Western operas.

( ) 4. Mr. Clive is not sure if he can understand the opera.

( ) 5. Mr. Clive drops a hint that he will purchase products from Andrew Johnson's company.

Task 5

Jessica and Andrew Johnson are talking about CAM. Listen to the conversation and answer the following questions.

1. What is CAD?

   _____

2. How do CAD and CAM systems work together in manufacture?

   _____

3. What is the main function of CAM?

   _____

4. What is the manufacturing process controlled by in a CAM system?

   _____

5. What are the advantages of CAM mentioned in the talk? Please note down one of them.

   _____

Speaking

Task 1 Work in pairs. Practice making short conversations with the words provided according to the example below.

> Example: Production Department/Mr. Brown/Bob/have trouble operating the NC machine tool/in the third workshop

Mr. Brown: Production Department.
Bob: Is that you Mr. Brown? This is Bob.
Mr. Brown: Oh hello, Bob. Nice to hear you.
Bob: I'm calling to ask for your help. I have trouble operating the NC machine tool. Would you please kindly come to help?
Mr. Brown: Sure. But I'm having a meeting this afternoon. How about tomorrow morning?
Bob: Let me see… tomorrow morning is Wednesday. OK, let's make a deal. I'll be waiting for you in the Third Workshop.
Mr. Brown: OK. See you then.
Bob: See you.

1. Production Department/Mr. King/Mike/discuss a production plan/in the meeting room
2. Marketing Department/Mr. Black/Wang Ning/discuss a marketing plan/in Mr. Black's office
3. ABC Company/Mr. Jones/Jane/Jane invites Mr. Jones to have dinner with her/pick him up at 6 pm

Task 2 Work in pairs. Bob waited for Mr. Brown from 8:00 to 12:00 on Wednesday morning, but Mr. Brown didn't appear. Bob was disappointed and decided to call Mr. Brown. Practice making a conversation. You may use the phrases or expressions listed below.

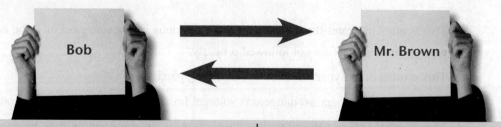

Bob	Mr. Brown
I have a hard time getting through to you.	I was listening to the lecture…
Did you forget…?	It was completely out of my mind.
You stood me up.	I attended the lecture…
I waited for you from… to…, but you didn't show up.	I'm awfully sorry.
I'd like…	I'll treat you to dinner.

English for Mechanical & Electrical Engineering

Task 3 Work in pairs. Bob is having an on-the-job training program in the evening school. He is discussing CAM with Mr. Brown. Make a conversation according to the instructions below.

Mr. Brown

Ask Bob how the evening training course is going.

Ask Bob what he has learnt.

Tell Bob that CAM is used in our company in three main areas.

Tell Bob that productivity and efficiency has been greatly raised with CAM.

Bob

Tell Mr. Brown you have learnt a lot.

Describe briefly to Mr. Brown what CAD/CAM is.

Ask about the application of CAM in our company.

Ask about the advantages of applying CAM.

Task 4 Work in pairs. Work with a partner to match each symbol with its indication. Then draw any of the symbols on a piece of paper and take turns to tell your partner its meaning and where it can be seen.

1. This symbol indicates that lethal accidents or serious injuries may occur if the operating and working instructions are not followed properly.

2. This symbol draws your attention to something particular.

3. This symbol warns against dangerous voltage! Immediate death might be the consequence.

4. This symbol informs the user that the system or its components may be damaged if the working and operating instructions are not followed.

5. This symbol indicates compulsory actions that must be performed by the operator.

6. This symbol indicates prohibitive actions that must not be performed by the operator.

Reading B

Plastic Molding

Plastics can be molded into various forms and hardened for commercial use. Plastic molding products can be seen everywhere. Examples are jars, protective caps, plastic tubes, toys, bottles, cases, accessories, kitchen utensils and a lot more. The keyboard and the mouse that you use are made through plastic molding and even the plastic parts of the chair that you are sitting on are created this way.

The basic idea in plastic molding is inserting molten liquid plastic into a ready shaped mold, for example, the mold of a bottle. It will be then allowed to cool, then the mold will be removed to reveal the plastic bottle.

If you are planning to go into the plastic molding business, you should first know the different processes. Here are basic definitions of various methods of plastic molding.

Injection Molding

In injection molding, melted plastic is forced into a mold cavity. Once cooled, the mold can be removed. This plastic molding process is commonly used in mass-production of a product. Injection molding machines were made in the 1930's. These can be used to mass produce toys, kitchen utensils, bottle caps, and cell phone stands.

Blow Molding

Blow molding is like injection molding except that hot liquid plastic pours out of a barrel vertically in a molten tube. The mold closes on it and forces it outward to conform to the inside shape of the mold. When it is cooled, the hollow part is formed. Examples of blow molding products are bottles, tubes and containers.

Compression Molding

In this type of plastic molding, hard plastic is pressed between two heated mold halves. Compression molding usually uses vertical presses instead of the horizontal presses used for injection and blow molding. The parts formed are then air-cooled.

Rotational Molding

Hollow molds packed with powdered plastic are secured to pipe-like spokes that extend from a central hub. The hub swings the whole mold into a closed furnace room causing the powder to melt and stick to the insides of the tools. As the molds turn slowly, the tools move into a cooling room. Here, sprayed water causes the plastic to harden into a hollow part.

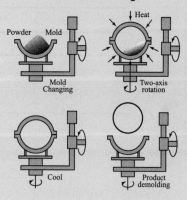

Rotational molding

Task 1 Read the passage and discuss the following questions.

1. In addition to the items mentioned in the text, can you make some other examples of plastic molding products in your daily life?
2. What did the invention of injection molding machines mean to the manufacturer?
3. What is the difference between injection molding and blow molding?
4. What is the difference in terms of raw material among the four types of plastic molding?

Task 2 Match the English expressions with their Chinese meanings.

1. kitchen utensils
2. injection molding
3. mold cavity
4. blow molding
5. compression molding
6. horizontal press
7. vertical press
8. rotational molding

a. 吹塑成型
b. 立式压力机
c. 卧式压力机
d. 注塑成型
e. 滚塑成型
f. 模具型腔
g. 压缩成型
h. 厨具

Task 3 Translate the following paragraph into Chinese.

Hollow molds packed with powdered plastic are secured to pipe-like spokes that extend from a central hub. The hub swings the whole mold to a closed furnace room causing the powder to melt and stick to the insides of the tools. As the molds turn slowly, the tools move into a cooling room. Here, sprayed water causes the plastic to harden into a hollow part.

Unit 4 Production

Task 1 A lecture, titled Digital Mold and Literate Activity, will be given by Professor Eric Jackson at 8 am on Tuesday April 14 in the 9th Meeting Room of Blue-sky Mold. Try to write a notice to inform all the staff of the report.

Sample:

A meeting to deliver a work report for 2019 by the Principal is to be held at 2 pm Thursday, May 28, in the auditorium.

All students and faculty are requested to be present on time.

Principal's Office
Mar. 22, 2020

Notice

Task 2 Read the following table and then complete the Daily Production Report (DPR).

Product	Specification	Standard time	Actual time	Date of producing	Output	Defective items	Operator
Carburetor (化油器)	361v	45S	45S	4-6-2020	640	1	Jeff
Radiator (散热器)	FH-201	45S	45S	4-6-2020	640	1	Bob

A Daily Production Report in Die-casting Workshop

Date: 4-6-2020
Prepared by: Peter Stein
Unit production manager: Norman Brown
Examiner: Andrew Johnson

　　There are _____
(生产出640件型号为361V汽车化油器和640件型号为FH-201摩托车散热器) in the die-casting workshop on April 6, 2020. _____
_____ (每件标准工时均为45秒), and so is their actual time.
_____(包括2件残次品在内, 今日的总产量是1,280). Jeff is in charge of carburetor's production, and _____ (鲍勃负责摩托车散热器的生产).

English for Mechanical & Electrical Engineering

Project

Project Guidelines

This project aims to go through the procedures of production after product designing. The whole process is divided into four steps. Step One is about preparation of the production. Step Two is about making a production plan. Step Three is about the daily production. Step Four is about role-playing a dialog.

Please follow the *Task Description* to complete the project.

Task Description

1 Step One
- Set up a group of 4-6 students in your class;
- Decide on the product you are going to produce;
- Discuss with your group members how to purchase raw materials for the production; you may search online for related information.

2 Step Two
- Discuss with your group members what CAM will help in this production process;
- Make a production plan arrangement.

3 Step Three
- Divide your group into two sides;
- Discuss within your side something important or necessary about technical specifications and programming and then exchange your ideas with the other side.

4 Step Four
- Divide your group into two sides: one side being the unit production manager and the examiner, the other being the workers;
- Take turns to play each role: the side of the workers reporting the daily production to the manager, and the side of the manager and the examiner asking about the production.

⚙ Self-evaluation

Rate your progress in this unit.	D	M	P	F*
I can describe the functions and application of CAM.	☐	☐	☐	☐
I can understand the engineer's instructions.	☐	☐	☐	☐
I can discuss with a technician about product making.	☐	☐	☐	☐
I can introduce the production status to guests.	☐	☐	☐	☐
I can read a production report.	☐	☐	☐	☐
I can write a notice and a DPR.	☐	☐	☐	☐

***Note**: Distinction, Merit, Pass, Fail

New Words and Expressions

Reading A

New Words

aircraft /'ɛr,kræft/ n. 飞机，飞行器
assistance /ə'sɪstəns/ n. 帮助，援助
coded /'kodɪd/ a. 编码的
coordinate /ko'ɔrdn,et/ v. 协调
database /'detə,bes/ n. 数据库
drill /drɪl/ v. 钻孔
essentially /ɪ'sɛnʃəlɪ/ ad. 本质上，根本地
extract /ɪk'strækt/ v. 摘出，选取
generality /,dʒɛnə'rælətɪ/ n. 通性，普遍（性）
interface /'ɪntəfes/ n. 接口，接合处
interpret /ɪn'tɜprɪt/ v. 解释，翻译
inventory /'ɪnvən,tɔrɪ/ n. 库存
load /lod/ v. 加载，装入程序
machine /mə'ʃin/ v. 以机器制造
numerical /nu'mɛrɪkl/ a. 数字的
perimeter /pə'rɪmətə/ n. 周边，周长
previously /'privɪəslɪ/ ad. 以前，先前
review /rɪ'vju/ v. 审查，回顾
sequence /'sikwəns/ n. 系，一连串
tolerance /'tɑlərəns/ n. （偏离标准的）容许误差，公差

Phrases & Expressions

be defined as 被定义为
center around 以……为中心，围绕
rely on 依赖，依靠

Reading B

New Words

barrel /'bærəl/ n. 圆筒，筒状物
cavity /'kævətɪ/ n. 腔，凹处
container /kən'tenə/ n. 容器
furnace /'fɜnɪs/ n. 熔炉
harden /'hɑrdn/ v. 变硬，凝固
hollow /'hɑlo/ a. 中空的
hub /hʌb/ n. 轮轴，轮毂
jar /dʒɑr/ n. 瓶，罐
mass /mæs/ n. 大量
molten /'moltn/ n. 熔化的，熔融状的
outward /'aʊtwəd/ ad. 向外
powder /'paʊdə/ n. 粉末
reveal /rɪ'vil/ v. 显露出，展现出
rotational /ro'teʃənəl/ a. 旋转的
spoke /spok/ n. 轮辐
spray /spre/ v. 喷洒，喷射
swing /swɪŋ/ v. 使旋转，摆动
tube /tub/ n. 管，管状物
utensil /ju'tɛnsl/ n. 器皿，用具
vertically /'vɜtɪklɪ/ ad. 垂直地

Phrases & Expressions

conform to 和……一致

Vocabulary and Structure

Task 1 Write out the words in Reading A or Reading B according to their meanings in the right column. The first letters are already given.

r_____ the capacity to withstand something

i_____ to make or be made into a whole

s_____ possessing or displaying symmetry

a_____ the act of assisting

c_____ to try to be more successful than someone or something else

n_____ of, relating to, or denoting a number or numbers

i_____ to put liquid, especially a drug, into someone's body by using a special needle

r_____ to trust or depend on someone or something

Task 2 Fill in each blank with the appropriate form of the word given in the brackets.

1. There wasn't a very (define) _____ reason why she should do so.

2. If goods are not well made you should complain to the (manufacture) _____.

3. The (apply) _____ of what you know will help you solve new problems.

4. I have been given an (inject) _____ by the doctor.

5. A healthy person offers more (resist) _____ to disease than a weak person.

6. He said he was in great need of my (assist) _____.

7. Cooking times may (various) _____ slightly, depending on your oven.

8. It is difficult to get (accuracy) _____ figures on population numbers.

9. They (replace) _____ the old machines with new ones.

10. Such rude behavior is not (tolerate) _____.

Task 3   Complete the following sentences with the words or phrases given below. Change the form if necessary.

| avail | lead to | involve with | in addition to | competitive |
| coordinate | link to | secure | rely | conform to |

1. Traditional booksellers face stiff _____ from companies selling via the Internet.
2. What one thinks and feels is mainly _____ tradition, habit and education.
3. Peter does not _____ the stereotype of a policeman.
4. There were no tickets _____ for Friday's performance.
5. _____ such subjects, the department also taught mathematics and geography.
6. We need good hand-eye _____ to play the ball game.
7. She's been _____ animal rights for many years.
8. She had proved that she could be _____ in a crisis.
9. The gate won't stay open, so we'll have to _____ it to that post.
10. Too much work and too little rest often _____ illness.

Task 4   Make sentences with the same pattern as is shown in the examples.

Example 1:   the use of computers as an aid to process planning is comparatively recent and has led to a rebirth of/what is known as group technology (GT)
— The use of computers as an aid to process planning is comparatively recent and has led to a rebirth of <u>what is known as group technology (GT)</u>.

Example 2:   the pictures are vivid and descriptive: they can show/you/what does the ancient city look like
— The pictures are vivid and descriptive: they can show you <u>what the ancient city looks like</u>.

1. he wants to make sure/when will Tom be here

2. this depends on/how hard do you work

3. our success is totally dependent on/whether would she give us some help

4. she had to tell me/what had happened to her by that day

5. people have heard/what has the President said/they are waiting to see/what will he do

Example 1: hurry up/you'll miss the train
— Hurry up, *otherwise* you'll miss the train.
Example 2: leave the horse alone/it would kick you
— Leave the horse alone, *otherwise* it would kick you.

1. he must be ill/he is present

2. the homework must be well done/our teacher will criticize us

3. take the opportunity/you will regret it

4. heat the water/it will freeze

5. she was out/I should have seen her

Task 5   **Translate the following sentences into English using the words or phrases given in the brackets.**

1. This technology is moving _____ (正朝着更广泛的应用方向发展). (in the direction of)

2. CAD/CAM will _____ (为未来的计算机集成工厂提供技术基础). (provide for)

3. The discussion _____ (以减少浪费为中心). (center around)

4. The class _____ (年龄从15岁到18岁不等). (vary from... to)

5. This computer _____ (并没有与制造过程直接连在一起). (be linked to)

Grammar

Sentences

Task 1   Put a "√" on the line if the corresponding expression is a sentence.

1. The sun rises in the east. _____√_____
2. To write a letter this evening. _____
3. Working together to save our environment. _____
4. The food smells delicious. _____
5. Too much homework to finish before class. _____
6. He hopes to fly to the moon. _____
7. A story with deep thoughts and emotions. _____
8. He gave me a pen. _____
9. Such as electrical, chemical and industrial engineering. _____
10. He heard somebody knocking on the window. _____

Task 2   Decide whether the following sentences are Simple, Compound or Complex.

1. Sara began planning her summer vacation in December. ___Simple___
2. Because I left the play early, I missed the surprise ending. _____
3. Tanya was invited to a party, so she wanted to buy a new dress. _____
4. Because of rain the baseball game was postponed. _____
5. English is not easy, but I like it very much. _____
6. Duane didn't pass the test, although he studied hard last week. _____
7. They won the match last year and wanted to win it again. _____
8. She sold her house, yet she can't help regretting it. _____
9. He is the man I saw in the park yesterday. _____
10. Stepping carelessly off the pavement, he was knocked down by a bus. _____

Task 3   Combine the following pairs of simple sentences into compound sentences using the words in the brackets.

1. We will go out to dinner tonight. They will join us. (and)
   We will go out to dinner tonight and they will join us.
2. I would like to get this job done in a hurry. I think it will take a long time. (but)
   _____
3. We could have the meeting tomorrow. We may postpone it until next Monday. (or)
   _____
4. We didn't choose the first class seat. We had to save money. (for)
   _____
5. I can't study everything all at once. I will study the most important concepts. (so)
   _____
6. I don't enjoy the study of chemistry much. I don't like the other natural sciences. (nor)
   _____
7. Lily was a successful career woman. Her husband wanted her to be a housewife. (yet)
   _____
8. Be quick. We'll be late for class. (or)
   _____
9. Come a little earlier next time. You'll miss the best part of the TV show. (or)
   _____
10. The teacher told them to clean the lecture room. He quickly walked away. (but)
    _____

Task 4   Underline each dependent clause in the following passage. If the dependent clause includes another dependent clause, underline it twice.

You ask *how I met my boyfriend*. Well, it's quite a funny story. Do you remember I failed one of my final exams? That meant I had to spend part of the summer in college. And that meant I couldn't go on holiday with my family. The travel company refused to give us a refund because we canceled too late. I was pretty angry about it. Then something nice happened. I think the travel agent felt sorry for me, because he had failed his final exam when he was a student. He agreed to transfer my booking to another tour which started later in the summer. I was really pleased. My father was too, as transferring the booking meant that his money wasn't being wasted. So, I went on this tour. And I met this young man. He was on his own too. We were the only ones traveling alone, so we found ourselves going around the sights together. He hadn't read about the places we were visiting and I spent most of my time telling him about them. We found we'd fallen in love at the end of the tour.

英语中的句子

根据句子的结构，英语中的句子可分为简单句（simple sentence）、并列句（compound sentence）和复合句（complex sentence）。

简单句：含一个主语（或并列主语）和谓语（或并列谓语）的句子。基本句型（pattern）包括：
1. 主语+系动词+表语（SVP）：He is a student.
2. 主语+不及物动词（SV）：My tooth aches.
3. 主语+及物动词+宾语（SVO）：Henry bought a dictionary.
4. 主语+及物动词+双宾语（SVOO）：My father bought me a car.
5. 主语+及物动词+宾语+宾补（SVOC）：Tom made the baby laugh.

注：其他各种简单句都可由这五种基本句型扩展、变化或省略而成。

并列句：由并列连词（coordinator）把两个或两个以上的简单句合并而成的句子。并列连词主要有：and, but, so, for, or, nor, yet。例如：
1. I asked her to have dinner together, but she was too busy.
2. He studied hard, yet he failed.

复合句：含有一个或一个以上从句的句子。从句一般由从属连接词（subordinator）、关系代词（relative pronoun）或关系副词（relative adverb）等引导，这些从句包括名词性从句（主语从句、宾语从句、表语从句及同位语从句）、定语从句和状语从句。例如：
1. Who will be our monitor hasn't been decided yet. (主语从句)
2. I am interested in what she is doing. (宾语从句)
3. This is what we should do. (表语从句)
4. He made a promise that he would never come late. (同位语从句)
5. He is the man who wants to see you. (定语从句)
6. If you are not too tired, let's go out for a walk. (状语从句)

Comprehensive Exercises

Task 1 Make sentences with the words and tips given.

1. the local school/attends/my son Tim
   SVO
2. to his school/my wife and I went/yesterday
   SV
3. we/to his teachers/spoke
   SV
4. Tim's school report/the teachers/gave us
   SVOO
5. very good/wasn't/Tim's report
   SVP
6. in every subject/were/his marks/low
   SVP
7. made him/Tim's report/very anxious
   SVOC
8. to try harder/my wife and I/told him
   SVOC
9. seems/Tim's friend Jimmy/very clever
   SVP
10. good marks/he got/in all subjects
    SVO

Task 2 Combine the following simple sentences into complex sentences with the words given in the brackets.

1. They didn't get the contract. He told us. (that)
2. They are the new secretaries. They work in our office. (who)
3. She is the nurse. I saw her at the hospital. (whom, that)
4. They are the children. Their football team won the match. (whose)
5. He had already opened the letter. He realized it wasn't addressed to him. (before)
6. I will give the letter to him. I see him. (as soon as)
7. I'll never forget the school. There I started to learn to play the violin. (where)
8. He didn't come to the meeting. He was injured in a car accident yesterday. (because)
9. It was such a big box. Nobody could move it. (that)
10. We will go to the beach. The weather is fine. (if)

Task 3 Fill in the blanks with appropriate coordinators, subordinators, relative pronouns or adverbs.

> Dear Sir,
>
>   I wish to complain about the villa holiday 1 _____ I booked with your company. The brochure described it as "in a peaceful setting", 2 _____ this was not the case. In fact, the villas were next to a building site, 3 _____ were noisy throughout the day.
>   4 _____ the villa was comfortable, one of the chairs was broken, 5 _____ the kitchen was dirty. We also expected a daily maid service, 6 _____ this did not happen. In fact, we only saw the maid twice during our holiday, 7 _____ her service was not good.
>   8 _____ the holiday did not live up to the promises 9 _____ you offered in the brochure, I think 10 _____ I'm entitled to compensation.
>   I look forward to hearing from you soon.
>
>   Yours faithfully,
>   Dr. Brown

Task 4 Combine some of the simple sentences in the following passage into compound or complex sentences where possible or necessary. Don't start too many sentences with the word "I".

I still remember the first day when I came to college. It was a sunny day. Everything seemed fresh to me. All the freshmen were excited. I was very excited, too. I had long dreamed of becoming a college student. Finally, my dream had come true. This was really a turning point in my life. I looked at the modern classroom buildings and the large library. I felt proud of my college. I knew that going to college would be a good opportunity for me to obtain a great deal of knowledge. The knowledge would be useful for my future career. But I knew that studying at college was a great challenge to me. I had to learn how to overcome the difficulties in my study and life at college. Anyway, I was determined to study hard. I had to live up to the expectations of my parents and my friends. I was sure that I would meet the challenge. I would make the best of the opportunity. I would prove myself a worthy college student.

_____
_____
_____
_____
_____
_____
_____
_____

Fun Time

Wrong Email Address

A man left the snowy streets of Chicago for a vacation in Florida. His wife was on a business trip and was planning to meet him there the next day. When he reached his hotel, he decided to send his wife a quick email. Unable to find the scrap of paper (记事纸片) on which he had written her email address, he did his best to type it in from memory. Unfortunately, he missed one letter, and his note was directed instead to an elderly preacher's wife (老牧师的妻子) whose husband had passed away only the day before. When the grieving (悲伤的) woman read her email, she let out a piercing scream (发出凄厉的尖叫声), and fell to the floor. At the sound, her family rushed into the room and saw this note on the screen:

"Dearest Wife,

Just got checked in (已登记入住). Everything prepared for your arrival tomorrow.

Your Loving Husband

P.S. Sure is hot down here."

UNIT 5

Product Inspection

Unit Objectives

After studying this unit, you are able to:
- know the difference between inspection and testing
- discuss inspection technologies with others
- consult other people about a certain inspection
- set up the procedure for a product inspection
- understand and keep inspection records

English for Mechanical & Electrical Engineering

Warming-up

Task 1 As a quality inspection engineer, Michael Button may experience the following situations. Match each situation with its corresponding picture.

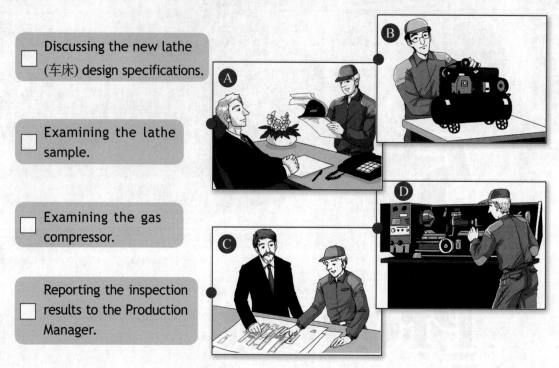

- ☐ Discussing the new lathe (车床) design specifications.
- ☐ Examining the lathe sample.
- ☐ Examining the gas compressor.
- ☐ Reporting the inspection results to the Production Manager.

Task 2 The following are what Michael Button usually does for quality inspection. Arrange them in the order of time and explain.

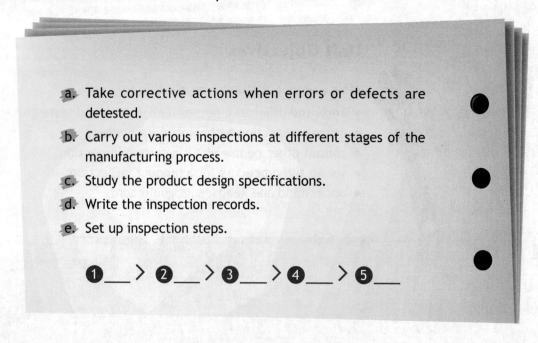

a. Take corrective actions when errors or defects are detested.
b. Carry out various inspections at different stages of the manufacturing process.
c. Study the product design specifications.
d. Write the inspection records.
e. Set up inspection steps.

❶___ > ❷___ > ❸___ > ❹___ > ❺___

Reading A

Task 1 Before reading the passage, try to answer the questions about product inspection.

1. When should product inspection be performed? What are the objectives of it?
2. Do you think product inspection is important to people's life? Did you ever buy things that are not to standard and how did you deal with them?

Inspection and Testing

Inspection and testing are industrial activities which ensure that manufactured products, components, and systems can meet their intended purpose and pick out defectives and missing or damaged ones. They are the operational parts of quality control which is most important to any manufacturing company.

Inspection and testing are performed before, during, and after manufacturing to ensure that the quality of the product is able to meet the design standards.

Whereas inspection is the activity of examining the product or its components to determine if they meet the design standards, testing is a procedure in which the item is observed during operation in order to determine whether it functions properly for a reasonable period of time.

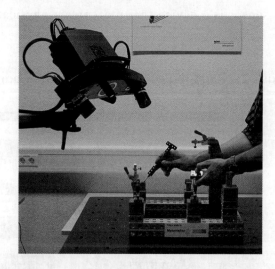

One of the most important methods to inspect the manufacturing process is statistical quality control. In statistical quality control, quality of a product is inferred from a sample taken from the population of the items. The sample of items is generated randomly from the population. Each item in the sample is inspected or tested for certain attributes and quality characteristics. The objective of statistical quality control is to determine when the process has gone out of control, so that corrective action can be taken.

The two principal techniques in statistical quality control are acceptance sampling and control charts. In acceptance sampling, a sample is taken from a batch of parts or products. Whether the batch will be accepted or rejected depends on the number of parts or products that pass the inspection or testing. A control chart is a simple graphical technique used to monitor and control a single characteristic of a manufacturing process. The objective is to estimate the principal parameter that describes the variability of this characteristic and then to use a test of hypothesis to determine if the process is in control.

To improve product quality and inspecting

quantity and changes it into a signal which can be read by an inspector or by an instrument. Sensor technologies for inspection can be divided into two types: contact and noncontact inspection methods. Contact inspection methods involve the use of a mechanical probe or other devices that make contact with the object being inspected. Noncontact inspection methods involve the use of a sensor located at a certain distance from the object to measure the desired features. Two important advantages of noncontact inspection methods are shorter inspection time and less damage to the object.

efficiency, sensor technology has been developed. A sensor is a device that measures a physical

Task 2 Read the passage. Match each paragraph with the information it contains.

Paragraph 1	statistical quality control and its objective two
Paragraph 2	types of sensor technologies
Paragraph 3	two main techniques of statistical quality control
Paragraph 4	time to perform inspection and testing
Paragraph 5	difference between inspection and testing
Paragraph 6	importance of inspection and testing

Task 3 Read the passage again and decide whether the following statements are true (T) or false (F).

☐ 1. Inspection and testing are part of the operational techniques and activities to assure the quality of a product.

☐ 2. Quality control is very important to the survival of a manufacturing company.

☐ 3. Inspection and testing are only performed after the product is manufactured.

☐ 4. The objective of testing is to examine whether a product or its components meet its design standards.

☐ 5. The procedure of determining whether a product functions properly during operation is called inspection.

☐ 6. In statistical quality control, the sample of products is chosen by chance rather than according to a plan.

☐ 7. Most of the items in the sample are inspected or tested for certain quality characteristics.

☐ 8. Corrective measures will be taken when the manufacturing process is found to be out of control.

☐ 9. Whether a batch of items will be accepted or rejected depends on the number of items that pass the inspection or testing.

☐ 10. Acceptance sampling is a graphical technique to control a single characteristic of the manufacturing process.

Task 4 There are two types of sensor technologies for inspection: contact and noncontact inspection methods. What's the difference between them? Discuss it with your partner.

Listening

Task 1 Michael is answering a call from David. Listen to the conversation and match the people with the correct information.

- the Production Manager
- a designing engineer
- notified of having a meeting
- will discuss product design with Michael
- works in Quality Inspection Department

English for Mechanical & Electrical Engineering

Task 2 Michael is talking about 1236SD Lathe with Mariah at the meeting. Listen to the conversation and fill in the blanks with what you hear.

Michael: Mariah, could you tell me the most striking 1 _____ of 1236SD Lathe? To carry out inspection, we need to have a broad knowledge of the product to be manufactured and its 2 _____ use.

Mariah: Well, different from other products, the 1236SD Lathe is designed to enable the user to take a 3 _____ position when they operate it.

Michael: So SD 4 _____ for Sit Down.

Mariah: Right. And it will be built with the same strong 5 _____ as our full-size lathes.

Michael: Yes.

Mariah: A high-quality, precision 6 _____, precision machined bedways (床身导轨), as well as our 7 _____ locking mechanism are all standard features of this machine.

Michael: Spindle, bedways and locking mechanism…

Mariah: Here is the detailed designing 8 _____.

Michael: I will take it back to the Inspection Department to study it together with my 9 _____ before we can set up a sequence of 10 _____.

Mariah: Contact me anytime you have a problem.

Task 3 Michael is discussing an inspection scheme with Monica, the Quality Inspection Manager. Listen to the conversation and choose the best answer to each question.

1. Which inspection will be conducted first?
   A. First Piece Inspection.
   B. Receiving Inspection.
   C. Batch Inspection.
   D. Sampling Inspection.

2. According to Monica, which inspection is of great importance?
   A. First Piece Inspection.   B. Receiving Inspection.
   C. Final Inspection.   D. Sampling Inspection.

3. How often will they carry out the sampling inspections?
   A. Twice a week.   B. At set intervals.
   C. At random.   D. Every day.

4. When is the Final Inspection usually carried out?
   A. Before the lathes are ready for packaging.
   B. After the lathes are transported to the customers.
   C. Before the lathes are installed for a testing.
   D. When the material and the parts from the suppliers arrive.

5. What does Monica remind Michael to do at last?
   A. To send her the inspection scheme.
   B. To check missing parts of the lathes.
   C. To study the designing specifications.
   D. To keep records of the inspections.

New Words

| First Piece Inspection | 首件检查 |
| at set intervals | 定期地 |

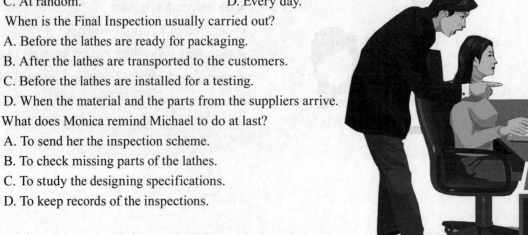

Unit 5 Product Inspection

Task 4 Diana, a new assistant in Quality Inspection Department is asking Michael about First Piece Inspection. Listen to the conversation and answer the following questions.

New Words
authorize v.   批准，委任
dimension n.   尺寸

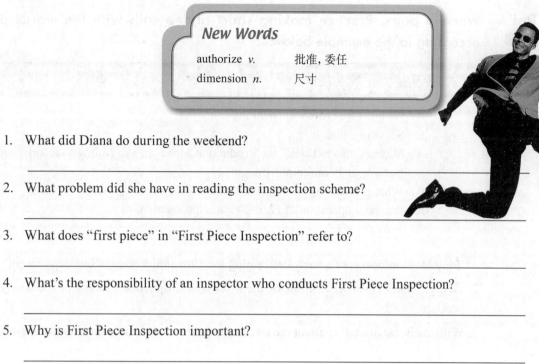

1. What did Diana do during the weekend?
   _____

2. What problem did she have in reading the inspection scheme?
   _____

3. What does "first piece" in "First Piece Inspection" refer to?
   _____

4. What's the responsibility of an inspector who conducts First Piece Inspection?
   _____

5. Why is First Piece Inspection important?
   _____

Task 5 Monica is calling Michael. Listen to the conversation and take notes of the information you hear.

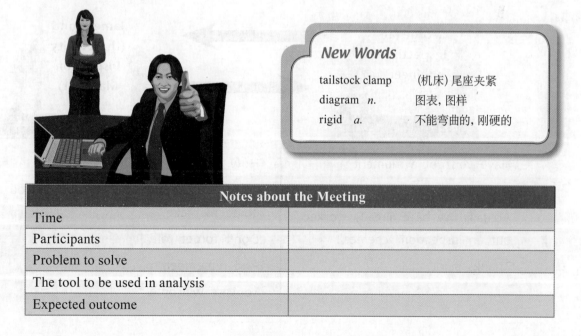

New Words
tailstock clamp   （机床）尾座夹紧
diagram n.   图表，图样
rigid a.   不能弯曲的，刚硬的

Notes about the Meeting	
Time	
Participants	
Problem to solve	
The tool to be used in analysis	
Expected outcome	

Speaking

Task 1 Work in pairs. Practice making short phone calls with the words provided according to the example below.

> Example: Michael/David/inform... of a meeting/preparation of 1236SD Lathe production

> A: Hello! Is that Michael speaking?
> B: Speaking, please.
> A: Michael, this is David, the Production Manager. I'm calling to inform you of the meeting at 10 am tomorrow.
> B: What is it for?
> A: For the preparation of 1236SD Lathe production.

1. Tony/Mess/inform... of a party/celebrating the 10th anniversary of our department
2. George/James Bond/let... know of the discussion/inspection process of the newly designed product
3. William/Beckham/tell... about the training courses/statistical quality control methods

Task 2 Work in pairs. Suppose George is seeking advice from James Bond about an inspection scheme on a newly designed lathe. Practice making a conversation. You may use the expressions listed below.

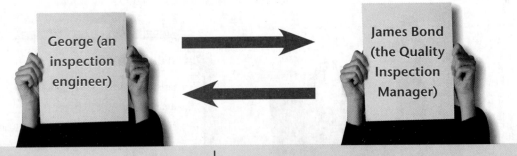

George (an inspection engineer)	James Bond (the Quality Inspection Manager)
Have you got a moment? I'm working on... Actually, we have already worked out an inspection scheme... Thanks a lot.	How is it going? Could you tell me what inspections you are going to make? Don't forget to... And remember to... You are welcome.

Unit 5 Product Inspection

Task 3 Work in pairs. Suppose Lily is a new assistant in Quality Inspection Department. She is asking George about First Piece Inspection. Make a conversation according to the instructions below.

Lily	George
Greet.	Greet. Ask how Lily spent the weekend.
Tell him you have difficulty in understanding the inspection scheme.	Offer help.
Accept the offer and ask him what "first piece" in the First Piece Inspection means.	Explain.
Ask whether the First Piece Inspection is important.	Show her the importance of First Piece Inspection.
Express thanks.	Respond to thanks.

Task 4 James is calling George, discussing with him about the meeting to be held in the afternoon. Make a conversation with your partner with the help of the instructions below.

James	George
○ Ask the receiver to get George on the phone.	○ Tell him you know what the meeting is for.
○ Ask George whether he knows the meeting to be held.	○ Tell him the problem found in the First Piece Inspection.
○ Ask him what the problem is.	○ Express the hope to take some corrective measures.
○ Ask whether he has noted down the measurements and observations on the inspection report.	○ Say you will use some tools that help analyze the complex problem.

Reading B

Inspection Record of a Precision Bench Lathe

Keeping inspection records is a key element in conducting inspections. The primary purpose of a record is to communicate and document the findings of an inspection. It is intended to include the important characteristics to be checked and all the essential inspection details. The chart below is an Inspection Record of a Precision Bench Lathe.

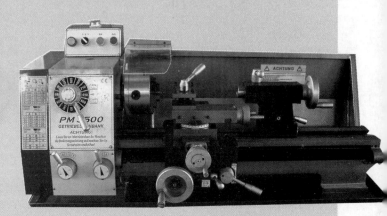

MODEL	G4003
INSPECTION DATE	June 5, 2020
INSPECTOR	Michael Button
PRODUCTION MANAGER	David Willington

	TEST	MEASURED VALUE (Unit: MM)
1	Tailstock Way Alignment	0.016
2	Spindle Center Runout	0.01
3	Spindle Nose Runout	0.01
4	Headstock Alignment Vertical	0.015
5	Headstock Alignment Horizontal	0.01
6	Tailstock Spindle Alignment Horizontal	0.01
7	Tailstock Spindle Alignment Vertical	0.021
8	Vertical Alignment of Head and Tail Centers	0.05
9	Face Plate Runout	0.02
		0.05
10	Chuck Runout	0.05
		0.05
		0.06
Remarks	All features are in accordance with the designing specifications.	

Task 1   Read the passage and answer the following questions.

1. What is the main purpose of keeping inspection records?
2. What does an inspection record usually contain?
3. When is the inspection of Precision Bench Lathe G4003 conducted?
4. Who conducted this inspection?
5. What's the result of this inspection?

Task 2   Match the English expressions with their Chinese meanings.

1. Spindle Center Runout
2. Spindle Nose Runout
3. Headstock Alignment Vertical
4. Headstock Alignment Horizontal
5. Tailstock Spindle Alignment Horizontal
6. Vertical Alignment of Head and Tail Centers
7. Face Plate Runout
8. Chuck Runout

a. 主轴端偏摆
b. 头架垂直校准
c. 划线平板偏摆
d. 头尾中心垂直校准
e. 主轴中心偏摆
f. 卡盘偏摆
g. 头架平行校准
h. 尾架主轴平行校准

Task 3   Translate the following paragraph in Reading A into Chinese.

> One of the most important methods to inspect the manufacturing process is statistical quality control. In statistical quality control, quality of a product is inferred from a sample taken from the population of the items. The sample of items is generated randomly from the population. Each item in the sample is inspected or tested for certain attributes and quality characteristics. The objective of statistical quality control is to determine when the process has gone out of control, so that corrective action can be taken.

English for Mechanical & Electrical Engineering

Writing

Task 1 Suppose you are Michael. Write a fax which is going to be sent to Monica, the Quality Inspection Manager who is attending a conference in another city. Tell her about the inspection you have made on the Precision Bench Lathe G4003. You should inform her of the inspection time and its result.

To:	Monica		
Phone:	893-7229	**Fax:**	893-7230
From:	Michael		
Phone:	671-6596	**Fax:**	671-6597
Date:	06/06/2020	**Pages:**	1
Cc:	David		
Subject:	Inspection on the Precision Bench Lathe G4003		

Dear Monica,

　　An _____ on the Precision Bench Lathe G4003 was _____ out yesterday. All of its main _____ were tested such as Tailstock Way Alignment, Spindle Center Runout and Spindle Nose Runout. They are all in _____ _____ with the design specifications. The inspection record was written, signed and kept in the Inspection Department. Further inspections will be following your directions.

　　　　　　　　　　　　　　　　　　　　　　　　　　　　_____,
　　　　　　　　　　　　　　　　　　　　　　　　　　　　　　　Michael

Task 2 Suppose you are Diana, an assistant in the Quality Inspection Department. Write a notice informing the staff members of the Quality Inspection Department to attend a meeting about the Inspection Scheme on 1236SD Lathe at 9 o'clock on Wednesday morning (June 10) in the small conference room.

Notice

　　　　　　　　　　　　　　　　　　　　　　　Quality Inspection Department
　　　　　　　　　　　　　　　　　　　　　　　　　　　　　June 8, 2020

⚙ Project

Project Guidelines

This project aims to go through the procedure of making a Finial Inspection of a lathe. The whole process is divided into three steps. Step One is to collect information about the designing specifications of a certain lathe. Step Two focuses on drafting the inspection scheme. Step Three rests on presenting the inspection scheme.

Please follow the *Task Description* to complete the project.

Task Description

① Step One

- Form an inspection group of 4-6 students in your class;
- Ask each member of the group to conduct a research about the designing specifications of a certain model of lathe. Use the library or the online resources;
- Collect your group members' information and make a brief presentation of the designing specifications of the lathe.

② Step Two

- Discuss with your group members about how to make the Final Inspection;
- Divide your group into two subgroups and ask each one to work out a scheme for the Final Inspection. You have to take into account the inspection time, equipments, methods, items to be tested, etc. when making the scheme.

③ Step Three

- Compare the two schemes and work out the Final Inspection Scheme to be used;
- Report it to the class and obtain some advice from your classmates.

Self-evaluation

Rate your progress in this unit.	D	M	P	F*
I can tell the difference between inspection and testing.	☐	☐	☐	☐
I know the general process of product inspection.	☐	☐	☐	☐
I can join in the discussion about inspection.	☐	☐	☐	☐
I can consult other people about inspection.	☐	☐	☐	☐
I can draft and keep inspection records.	☐	☐	☐	☐
I can write a fax.	☐	☐	☐	☐

*Note: Distinction, Merit, Pass, Fail

New Words and Expressions

Reading A

New Words

attribute /ə'trɪbjut/ n. 属性, 特性
batch /bætʃ/ n. 一批, 成批 (工作件)
defective /dɪ'fɛktɪv/ n. 残次品 a. 有缺陷的
estimate /'ɛstəmɪt/ v. 评估, 测量
graphical /'græfɪkl/ a. 绘成图表的, 图解形式的
hypothesis /haɪ'pɑθəsɪs/ n. 假设
infer /ɪn'fɝ/ v. 推断, 推论
inspection /ɪn'spɛkʃən/ n. 检查
intend /ɪn'tɛnd/ v. 计划, 打算
parameter /pə'ræmətɚ/ n. 参数, 参量
population /ˌpɑpjə'leʃən/ n. (统计学中) 母体, 总体
principal /'prɪnsəpl/ a. 主要的, 最重要的
randomly /'rændəmlɪ/ ad. 随机地
sensor /'sɛnsɚ/ n. 传感器
signal /'sɪgnl/ n. 信号, 指令

statistical /stə'tɪstɪkl/ a. 统计学的
variability /ˌvɛrɪə'bɪlətɪ/ n. 变量, 变率

Phrases & Expressions

be in control 在控制之中
go out of control 不受控制
pick out 挑出

Technical Terms

acceptance sampling 进料抽样试验
contact inspection methods 接触检测方法
control chart 管理图表
mechanical probe 机械探针
noncontact inspection methods 非接触检测方法
physical quantity 物理量
statistical quality control 统计质量管理
test of hypothesis 假设检测

Reading B

New Words

alignment /ə'laɪnmənt/ n. 校直, 调准
chuck /tʃʌk/ n. 卡盘
document /'dɑkjəmənt/ v. 用文件记录
finding /'faɪndɪŋ/ n. (pl.) 结果
headstock /'hɛdstɑk/ n. 头架
horizontal /ˌhɑrə'zɑntl/ a. 水平的
inspector /ɪn'spɛktɚ/ n. 检查员
primary /'praɪˌmɛrɪ/ a. 首要的, 主要的
remark /rɪ'mɑrk/ n. 评论, 备注
runout /'rʌnaʊt/ n. 偏斜, 偏心率
tailstock /'telstɑk/ n. 尾架, 尾座

vertical /'vɝtɪkl/ a. 垂直的

Phrases & Expressions

in accordance with 与……一致

Technical Terms

design specifications 设计规格
measured value 测量值
precision bench lathe 精密台式车床
spindle center 主轴中心
spindle nose 主轴端

Vocabulary and Structure

Task 1 Write out the words in Reading A according to their meanings in the right column. The first letters are already given.

e_____	to try to judge the value, size, speed, cost etc. of something, without calculating it exactly
f_____	to work or operate
h_____	an idea or explanation for something that is based on known facts but has not yet been proved
i_____	a tool or device used for performing a particular piece of work
o_____	something which you plan to do or achieve
p_____	(in statistics) any finite or infinite collection of items under consideration
s_____	a group of people or things that is chosen out of a larger number and is questioned or tested in order to obtain information about the larger group
v_____	the quality, state, or degree of being changeable

Task 2 Fill in each blank with the appropriate form of the word given in the brackets.

1. The plane appeared to have crashed because of a (machine) _____ problem.
2. The (graph) _____ results are entirely convincing.
3. She arrived to carry out a health and safety (inspect) _____ of the building.
4. From his manner, we drew the (infer) _____ that he was satisfied with the exam.
5. The machine won't function (proper) _____ if you don't oil it well.
6. The management took all (reason) _____ safety precautions.
7. (contact) _____ inspection is to use a device located at a certain distance from the product to examine it.
8. The country's power (generate) _____ is likely to drop further in April, indicating an economic recovery is still some way off.
9. Other researchers soon confirm their (find) _____.
10. The equipment must rest on (horizon) _____ and firm ground.

Unit 5 Product Inspection

Task 3 Complete the following sentences with the words or phrases given below. Change the form if necessary.

| a batch of | conduct | make contact with | details | document |
| feature | generate | go out of control | intend | randomly |

1. Fuel system converts hydrogen (氢气) to _____ electricity.
2. We are _____ a survey to find out what our customers think of their local bus service.
3. I've written the outline of my report, but I have to fill in the _____.
4. The car skidded and _____, crashing into an oncoming truck.
5. I _____ to catch the early train, but I didn't get up in time.
6. Never _____ select someone when you are lonely. That lacks responsibility.
7. _____ newly cooked homemade cakes was brought in by the cook.
8. The history of this area is very well _____.
9. We _____ the ship by radio.
10. Our latest model of phone has several new _____.

Task 4 Make sentences with the same pattern as is shown in the examples.

Example 1: inspection is the activity of examining the product or its components to determine if they meet the design standards/testing is a procedure in which the item is observed during operation in order to determine whether it functions properly for a reasonable period of time
— <u>Whereas</u> inspection is the activity of examining the product or its components to determine if they meet the design standards, testing is a procedure in which the item is observed during operation in order to determine whether it functions properly for a reasonable period of time.

Example 2: wise men love truth/fools shun (回避) it
— Wise men love truth, <u>whereas</u> fools shun it.

1. his children are well bred/those of his sister are naughty
   _____

2. you eat a massive plate of food for lunch/I have just a sandwich
   _____

3. some people like coffee/others like tea
   _____

4. he must be about 60/his wife looks about 30
   _____

5. we want a flat/they would rather live in a house
   _____

> *Example 1:* all features/accordance/the designing specifications
> — All features are *in accordance with* the designing specifications.
>
> *Example 2:* the goods/will be sent/accordance/your instructions
> — The goods will be sent *in accordance with* your instructions.

1. we/ship our goods/the terms of the contract
   _____

2. the rights of children/protect/the law
   _____

3. he/acted/his beliefs
   _____

4. article 47/may only be used/international law
   _____

5. use this product/only/the manufacturer's instructions
   _____

Task 5 Translate the following sentences into English using the words or phrases given in the brackets.

1. If you want to _____ (确保能赶上那班飞机), take a taxi. (ensure that)

2. Team-work _____ (是任何足球队成功的关键). (be essential to)

3. From your smile _____ (我们断定你很高兴). (infer that)

4. The study didn't _____ (没有包含任何人体试验). (involve)

5. I had a narrow escape yesterday when I had a puncture on the motorway, _____ (但是幸运的是，我能控制住车) until I could pull over and stop. (keep... in control)

Unit 5 Product Inspection

Grammar

Attributive Clauses

Task 1 Use the information in the brackets to modify the phrases.

1. the girl (she is pretty) → *the pretty girl*
2. the woman (she has long hair) → *the woman with long hair*
3. the man (he plays football) → *the man who plays football*
4. the woman (she is slim) →
5. the young boy (he is beside the window) →
6. the man (he drives a car) →
7. the boy (he has blue eyes) →
8. the young lady (she is smart) →
9. the boy (he failed his final exam) →
10. the girl (she takes evening class) →

Task 2 Complete the following sentences with appropriate relative pronouns or adverbs.

1. The story is about a girl <u>who/that</u> runs away from home.
2. The police have caught the men _____ stole my car.
3. A dictionary is a book _____ gives you the meaning of words.
4. It seems that the Earth is the only planet _____ can support life.
5. Have you finished the work _____ you have to do?
6. We stayed at the hotel _____ Peter recommended.
7. I recently went back to the town _____ I was born.
8. An orphan is a child _____ parents are dead.
9. Football players _____ take drugs will be banned from playing.
10. The person _____ the police were questioning has now been released.

Task 3 Complete the following conversation with appropriate relative pronouns or adverbs.

A: Shall we invite Carlo to this party?
B: Carlo? Who's he?
A: He's the Italian guy who is staying with John's family.
B: Oh, yeah. Is he the one 1 _whose_ wallet got stolen when they were in London?
A: That's right. They caught the guy 2 _____ did it, but he'd already spent all the money 3 _____ Carlo had brought with him.
B: Poor Carlo. Perhaps the party will cheer him up.
A: It might, if we ask the girl 4 _____ he's been going out with.
B: Who's that?
A: Delia's her name. She works in that cinema 5 _____ they show all the foreign films.
B: But will she be free on Thursday evening?
A: Yes, it's her evening off. That's the reason 6 _____ I suggested Thursday.
B: OK. Who else? What about Nicky and Cherry?
A: Are they the girls 7 _____ you went to France with?
B: Yes, if they bring their boyfriends, that'll be 10 of us. But have you got a room 8 _____ is big enough? My mother says we can't use our sitting-room because we made too much mess the last time 9 _____ she let us have a party.
A: It's all right. We've got a basement 10 _____ we store old furniture. If we clean it up, it'll be fine.
B: Great. Let's go and have a look at it.

Task 4 Underline the attributive clauses in the following passage.

One kind of vacation _that many Americans enjoy_ is camping. Each summer millions of Americans drive to the countryside where they find places to camp. The national parks, many of which are in the mountains, are favorite camping places. Campers enjoy the fresh air, the lakes and the forests which they find in these parks. Campers hike, swim and fish. They can also find many kinds of animals and plants in the parks. Mostly, campers relax. They enjoy a change from their busy lives in the city. Some campers have trailers which they drive or pull behind their cars to their camp sites. Trailers are like houses on wheels. They have many of the conveniences which people have in their homes, such as electricity and hot water. But most campers don't have trailers. They camp in tents which they set up on their camp sites. Campers in tents don't have the conveniences that campers in trailers have. Tent campers enjoy a very simple life.

定语从句

定语从句在句中作定语,修饰名词或代词,被修饰的名词或代词称为先行词 (antecedent)。定语从句通常出现在先行词之后,由关系代词 (relative pronoun) 或关系副词 (relative adverb) 引出。

1. 基本形式

关系代词	例 句	语法成分	功 能
who/that	Is he the man who/that wants to see you?	主语	指人
whom/that	He is the man (whom/that) I saw yesterday.	宾语 (可省略)	指人
whose	I met a man whose sister works in television.	定语	指人
whose	Please pass me the book whose cover is green.	定语	指物
which/that	I don't like stories which/that have unhappy endings.	主语	指物
which/that	I love the book (which/that) you lent to me.	宾语 (可省略)	指物
关系副词	例 句	语法成分	功 能
when	I remember the day when I met him.	状语	时间
where	The hotel where we stayed wasn't very clean.	状语	地点
why	The reason why I'm calling is to invite you to a party.	状语	原因

2. 限制性和非限制性定语从句

定语从句有限制性和非限制性两种。限制性定语从句是先行词不可缺少的部分,去掉它主句意思往往不明确;非限制性定语从句是先行词的附加说明,去掉了也不会影响主句的意思,它与主句之间通常用逗号分开。例如:

This is the house *which we bought last month*. (限制性)

The house, *which we bought last month*, is very nice. (非限制性)

3. 介词 + 关系词

1) 介词后面的关系词不能省略;

2) that 前不能有介词。

4. 只能用 that 而不用 which 作为定语从句的关系代词的情况

1) 先行词前有序数词或形容词最高级时;

2) 不定代词 anything, nothing, all, much, few, any, little 等作先行词时;

3) 先行词有 the only, the very, the one 修饰时;

4) 先行词既指人又指物时。

Comprehensive Exercises

Task 1 Fill in the blanks with *where, when, that* or *which*. Each blank may have more than one answer.

1. The city _____ I was born has a lot of lakes.
2. I don't like cities _____ have a lot of factories.
3. We like to shop at stores _____ have products from different countries.
4. I like to shop at stores _____ I can find products from different countries.
5. Mid-autumn Day is a time _____ family members meet together.
6. New Year's Eve is a time _____ I love.
7. Their anniversary is a date _____ has a lot of meaning for them.
8. My birthday is a day _____ I think about my past.
9. February is the only month _____ has fewer than 30 days.
10. Their vacation in London was the best time _____ they had ever had.

Task 2 Fill in each blank with *whom* or *which* plus a preposition.

1. The person _____ _____ I spoke just now is the manager that I told you about.
2. The pencil _____ _____ he was writing broke.
3. Wu Dong, _____ _____ I went to the concert, enjoyed it very much.
4. The two things _____ _____ Marx was not sure were grammar and some of the idioms of English.
5. Her bag, _____ _____ she put all her books, has not been found.
6. The stories about the Long March, _____ _____ this is one example, are well written.
7. In the dark street, there was not a single person _____ _____ he could turn for help.
8. The man _____ _____ I stayed won the shooting game.
9. The hospital _____ _____ I was born is seven kilometers away.
10. The person _____ _____ I borrowed a pen is Mr. Ball.

Task 3 Combine the following pairs of sentences into one sentence by changing the second sentence into an attributive clause.

1. The fan is on the desk. You want it.
   _____
2. The man is in the next room. He brought our textbooks yesterday.
   _____
3. The man was wearing a blue shirt. He witnessed the accident.
   _____
4. The students will not pass the exam. They don't study hard.
   _____
5. The woman is our geography teacher. You saw her in the park.
   _____
6. The letter is from my sister. I received it yesterday.
   _____
7. The man reported the accident. His car was damaged.
   _____

8. The research paper must be finished by Friday. David is working on it.

9. The church is very old. My grandparents were married there.

10. 1910 is the year. The revolution began then.

Find the mistake in each of the following sentences and correct it.

1. The bike I bought it yesterday was broken.

2. I know a girl speaks five languages.

3. The person to who you spoke is a friend of mine.

4. The house in that we live is very small.

5. A car which have a big engine is not economical.

6. This is the best film which I have ever seen.

7. Everything which we saw was of great interest.

8. His dog, that was now very old, became ill and died.

9. Who want to leave early should sit in the back.

10. He talked of things and persons whom he was interested in.

Fun Time

Westerners love humor which sometimes comes from the daily use of English language.

1

— What did a late tomato say to other tomatoes?

— I will ketchup (catch up).

英文中，番茄酱 (ketchup) 和赶上 (catch up) 发音相同，而该笑话的主角又恰好是一个迟到的番茄，使得谐音梗和语境完美契合，组成了英文中的冷笑话。

2

— Why shouldn't we give Elsa a balloon?

— Because she will Let It Go.

此处是一个与电影文化密切相关的梗。Elsa指代的是迪士尼电影《冰雪奇缘》的女主角，回答让我们又再次想到了Elsa那首耳熟能详的Let It Go，和语境契合度高，实现了幽默和文化输出。

3

"You're not eating your fish," the waitress asked him, "anything wrong with it?"

"Long time no sea," the customer replied.

鱼久不见海，肯定不新鲜。而Long time no see是朋友见面时所用的寒暄语，相当于老朋友久别重逢时说的"好久不见"。sea与see是同音异形异义词，这位顾客巧妙地运用这一点，指出了鱼不好吃，不新鲜。

UNIT 6

Installation and Maintenance

Unit Objectives

After studying this unit, you are able to:
- master some expressions about installation and maintenance
- describe the process of installing and cleaning a machine
- read a manual
- communicate with a client about maintenance
- write to a client

English for Mechanical & Electrical Engineering

Warming-up

Task 1 As a mechanic, Tom Waters may experience the following situations. Match each situation with its corresponding picture.

☐ Discussing with his co-workers about some technical problems.

☐ Demonstrating how to assemble a mini-lathe.

☐ Showing an intern student around the workshop.

☐ Keeping the tools clean and arranging them in the toolbox.

Task 2 The following things are what Tom usually does for installing a machine. Arrange them in the order of time and explain.

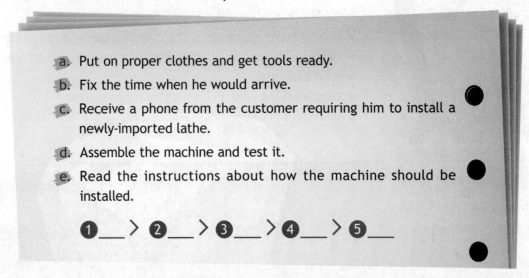

a. Put on proper clothes and get tools ready.
b. Fix the time when he would arrive.
c. Receive a phone from the customer requiring him to install a newly-imported lathe.
d. Assemble the machine and test it.
e. Read the instructions about how the machine should be installed.

❶___ > ❷___ > ❸___ > ❹___ > ❺___

Reading A

Task 1 Before reading the passage, try to answer the questions about machine installation.

1. How much do you know about a lathe? What is it used for in the machine manufacture?
2. Do you know the classifications of a lathe? Discuss with your classmates.

Mounting the Lathe to a Board

Mounting the lathe to a board (see Fig. 1) is necessary because of the narrow base. This keeps the machine from tipping. We recommend mounting the lathe on a piece of pre-finished shelf material, which is readily available at most hardware stores. The machine can be secured to the board using screws of a certain specification with washers and nuts. Lengths should be 1-1/2 inches" for short bed lathes and 1-7/8 inches" for long bed lathes. Rubber feet should be attached at each corner on the bottom of the mounting boards. They are also readily available in hardware stores. This arrangement gives the machines a stable platform for operation yet still allows for easy storage. The rubber feet help minimize the noise and vibration from the motor. Mounting the tool directly to the workbench can cause vibration of the bench itself, which acts as a "speaker" and actually amplifies the motor noise. Bench mounting also eliminates one of the best features of Sherline machines—the ability to be easily put away for storage. The mill may be mounted in a similar manner on a 10" x 12" inch to 12" x 24" inch pre-finished shelf board with rubber feet using the screws to attach the mill to the board.

REMEMBER: DO NOT LIFT YOUR MACHINE BY THE MOTOR! Carry the machine by lifting under the base or by the mounting board. To keep your Sherline tools clean, soft plastic dust covers are available. The lathe cover is P/N 4150 for the Model 4000/4100 and 4500/4530 short bed lathes and P/N 4151 for the Model 4400/4410 long bed lathe. A mill dust cover is available as P/N 5150 for 5000-series mills and P/N 5151 for 2000-series 8-direction mills.

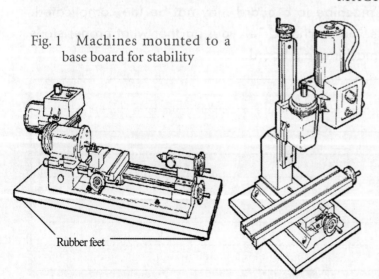

Fig. 1 Machines mounted to a base board for stability

English for Mechanical & Electrical Engineering

Task 2 Read the passage and match the actions on the left with their possible results on the right.

mounting a lathe to a board • • reducing the noise
attaching rubber feet • • achieving stability
mounting a lathe directly to the • • keeping the machine clean
workbench • causing the vibration of the
lifting a machine by the motor • machine
putting on a dust cover • • damaging the lathe

Task 3 Read the passage again and decide whether the following statements are true (T) or false (F).

☐ 1. The lathe is designed with a narrow base so that it will be easily transported.
☐ 2. The lathe is suggested to be fixed on a piece of board to amplify the noise.
☐ 3. The pre-finished shelf material can be easily obtained because they are available in hardware stores.
☐ 4. The rubber feet are attached to the bottom of the board in order to keep the machine stable.
☐ 5. Workbench may act as a "speaker" to talk with the workers.
☐ 6. The machine can never be lifted by the motor.
☐ 7. The machine can be secured to the board by some screws.
☐ 8. The mill can be mounted to a board in a similar way of mounting the lathe.
☐ 9. The vibration of the workbench may produce great noise in the workshop.
☐ 10. The plastic covers are used to keep the machine clean.

Task 4 Work in pairs. Mounting a machine to a board may not be too complicated. Suppose you are Tom Waters, the machanic. Tell your partner what proper steps should be taken to secure a machine to a board.

Unit 6 Installation and Maintenance

Listening

Task 1 Listen to the conversation and match the people with the correct information.

- has set off for Mr. Bird's company
- a mechanic of Darkin Machinery Company
- an intern of Darkin Machinery Company
- manager of the company that ordered the lathe
- will check the machine parts

Task 2 David Clinton, a mechanic from Dragon Machinery Co., Ltd, wants to know how to place an order for some spare parts from Darkin Machinery Company. Listen to the conversation and fill in the blanks with what you hear.

Jane: Good morning, this is Jane Peterson. Can I help you?

David: Morning, Miss Peterson. This is David Clinton, a 1 _____ from Dragon Machinery Co., Ltd in Chicago. I would like to know how to 2 _____ an order for some spare parts manufactured by your company.

Jane: Well, Mr. Clinton, first of all, you have to make sure the machine 3 _____ 4 _____ and the 5 _____ 6 _____ of the machine.

David: Okay, thank you for reminding me of those details.

Jane: Then you may fill in the form 7 _____ from our company's website, or you may also order those 8 _____ parts from the factory directly by 9 _____.

David: Great! What is the phone number?

Jane: Toll free order line 1-800-543-0756. Orders 10 _____ before 12 o'clock will be shipped the same day.

David: Thank you. Miss Peterson.

Jane: You are welcome. Goodbye!

David: Bye.

English for Mechanical & Electrical Engineering

Task 3 Listen to the conversation between Jane and Tom and choose the best answer to each question.

1. What does Jane want to know?
   A. How to repair a machine.
   B. How to disassemble the machine into parts.
   C. How to install a lathe.
   D. How to deliver a machine.
2. How many steps should Tom follow if he wants to install the machine?
   A. Three.   B. Four.   C. Two.   D. Five.
3. Which of the following is the second step to install the machine?
   A. Test the machine.
   B. Mounting the Z-axis onto the base.
   C. Installing the motor and speed control.
   D. Installing the cross-slide table.
4. Why should a machine be tested after it is assembled?
   A. Because the instructions tell us so.
   B. Because the mechanic experience tells us so.
   C. Because we want to know whether it can work properly.
   D. Because our intuition tells us to do so.
5. What can we infer from the dialogue?
   A. The machine is usually transported to the company in parts disassembled.
   B. After the machine has been installed, it can be put into use immediately.
   C. The machine installation generally follows four steps.
   D. Testing of the machine is very necessary to make sure that the machine can work properly.

Task 4 Jane Peterson wants to know something about the readouts. Listen to the conversation and answer the following questions.

1. Why do the digital readouts become more and more popular according to the conversation?
   _____

2. Does the user need any special knowledge to understand the readouts?
   _____

3. Are all the machines equipped with the same type of readouts?
   _____

4. Is the readout attached to the machine when it is transported to the factory?
   _____

5. What should a mechanic do if he does not want to lag behind the time?
   _____

Task 5 Tom Waters shows how to mount the cross-slide of a lathe to Jane Peterson. Listen to the short passage and take notes of the information you hear.

The installation of the cross-slide needs 1 _____.
Firstly, you have to make sure 2 _____ is covered with grease.
Secondly, make sure that the gib (凹字形楔) is 3 _____; if it has come off, you have to put it 4 _____. And lastly, set
5 _____ over the gib and match the dovetail
6 _____.

126

Unit 6 Installation and Maintenance

⚙ Speaking

Task 1 Work in pairs. Practice making short conversations with the words provided according to the example below.

Example: Hisense Group/John Wilson/signal/have a look

(John is knocking at the door; Mr. Smith opens the door.)

A: Good morning.
B: Good morning, are you Mr. Smith? I'm John Wilson from Hisense Group.
A: Mr. Wilson, come in, please. There is no signal shown on the TV screen.
B: Don't worry. Let me have a look.

1. Sichuan Changhong Electric Co., Ltd./Tim Smith/sound/have a look
2. Nanjing Panda Electronics Co., Ltd/Sam Adams/unclear pictures/replace some parts
3. Qingdao Haier Co. Ltd./waves on the screen /replace some parts

Task 2 Work in pairs. John is working in Hisense Group. He is now helping Mr. Smith to install the satellite dish (卫星接收器). Practice making a conversation. You may use the phrases or expressions listed below.

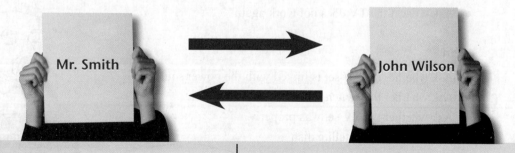

Would you please...?
Why should we choose a proper place?
How about the next step?
What if...?

There are several steps... Firstly, ...
Receive the signals better and prevent wind.
Connect the dish with the TV set.
Call this number for help.

Task 3   Work in pairs. John is receiving a telephone call from Mr. Smith, who says that the TV set can't receive any program. John promises to come to Mr. Smith's home and check what is wrong with the TV. Make a conversation according to the instructions below.

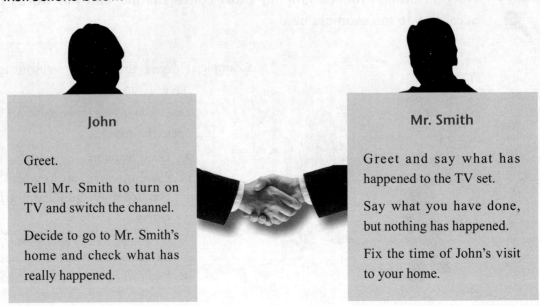

John

Greet.

Tell Mr. Smith to turn on TV and switch the channel.

Decide to go to Mr. Smith's home and check what has really happened.

Mr. Smith

Greet and say what has happened to the TV set.

Say what you have done, but nothing has happened.

Fix the time of John's visit to your home.

Task 4   Work in pairs. John is now at Mr. Smith's home and he is checking the TV set. Make a conversation with your partner with the help of the instructions below.

Mr. Smith

- Tell John when you found the TV set failed to work.
- Say that everything seems alright because no one touches it.
- Mention the storm the day before yesterday.
- Ask what if the TV does not work again.

John

- Ask whether the TV set is linked with the power source.
- Ask whether the switch of the TV set is in order.
- Ask whether the TV set was properly connected to the satellite dish.
- Find out the real reason and remind Mr. Smith to turn off the TV if there is a storm.

Reading B

Cleaning the Lathe

If a lathe is dirty or has not been in use for some time, its guide rails and spindles may need to be cleaned thoroughly.

Remove the covers from the X and Y guide rails and spindles. Spray plenty of anti-corrosion solution onto the spindles and guide shafts and let the solution sink in so that the old grease and oil residue can be eliminated.

Dried bits of dirt and old grease can then be removed using cleaning paper and/or a toothbrush. It is best to polish rusty areas with some cleaning paper and anti-corrosion solution. In extreme cases you can also use the polishing paste used to "shine" old cars. On no account must you clean the guide rails with a nail file or other hard instrument, as this is bound to damage them.

If unsuitable grease has been used the spindles may be covered in resin, in which case it is no longer possible to clean them with a solution and a toothbrush. Instead it is better to use a wooden wedge which is thin enough to fit in the spindle threads. Then you can remove the really tough dirt without damaging the spindles.

Depending on how bad the dirt is, the cleaning procedure will have to be repeated a number of times.

After that, grease the shafts again with a few drops of acid-free and resin-free sewing machine oil or multi-lube spray.

Finally the spindles are greased again with the aid of a clean brush. Make sure that the brush does not lose any hair and that the grease penetrates the grooves deeply.

Task 1   Read the passage and match each paragraph with the corresponding measure it describes to clean the lathe.

- Paragraph 1 •   • Clean the machine thoroughly.
- Paragraph 2 •   • Grease the shafts again.
- Paragraph 3 •   • Repeat the cleaning procedure.
- Paragraph 4 •   • Use a thin wooden wedge.
- Paragraph 5 •   • Grease the spindles.
- Paragraph 6 •   • Spray plenty of anti-corrosion solution.
- Paragraph 7 •   • Use cleaning paper or a toothbrush.

Task 2 Match the English expressions with their Chinese meanings.

1. anti-corrosion solution
2. the polishing paste
3. grease the shafts
4. hard instrument
5. tough dirt
6. sewing machine oil
7. multi-lube spray
8. cleaning procedure

a. 抛光膏，磨光剂
b. 抗腐蚀溶剂
c. 缝纫机油
d. 坚硬的器具
e. 顽固的灰尘
f. 清洁程序
g. 多功能润滑喷剂
h. 给机轴涂油

Task 3 Retell the measures taken to clean a lathe.

Task 4 Translate the following short paragraph into Chinese.

Anyone who is going to use Twister Speed Lathe needs to be aware of the following warnings:
1. Never attempt to operate this machine in a wet area.
2. Never leave this machine unattended while it is running.
3. Never wear gloves while operating this machine.
4. Don't operate this machine while wearing a ring, a watch, jewelry, a tie, loose clothing, or long hair which is not contained in a net, or a shop cap.

Unit 6 Installation and Maintenance

Writing

Task 1 Suppose you are Tom Waters. You are writing the cleaning instructions to Jane Peterson. Finish the instructions below according to Reading B. The first letter is already given.

1. R_____ the covers from the X and Y guide rail and s_____.
2. Spray plenty of anti-corrosion s_____ onto the s_____ and guide shafts.
3. Use cleaning paper or a t_____ to remove dirt and old g_____.

Task 2 Suppose you are John who visited Mr. Smith to fix his TV set last week. Write a follow-up letter to make sure that the TV set works properly and express your willingness to help in the future.

Dear Mr. Smith,

_____
_____
_____
_____
_____
_____
_____

Yours truly,
John Wilson
Maintenance Department of Hisense Group

Project

This project aims to learn the preventive maintenance of machines which is vital to a factory. The whole process is divided into three steps. Step One is to get the knowledge of preventive maintenance. Step Two focuses on comparing the different results between groups. Step Three is about household appliances.

Please follow the *Task Description* to complete the project.

Task Description

1 **Step One**
- Set up a group of 4-6 students in your class.
- Share the work of searching online for information about preventive maintenance of machines.
- Discuss with your group members about the importance of preventive maintenance and the aspects it concerns, and then write down results of the discussion.

2 **Step Two**
- Present the result to the other groups.
- Compare the results of each group.

3 **Step Three**
- Discuss the importance of preventive maintenance of home appliances.
- Divide your group into two: one side telling the other side how you take good care of the TV set or the refrigerator at your home, the other side putting forward questions, and then take turns to play each role.

🛠 Self-evaluation

Unit 6 Installation and Maintenance

Rate your progress in this unit.	D	M	P	F*
I can describe the process of installing a lathe to a board.	☐	☐	☐	☐
I can describe the process of cleaning a machine.	☐	☐	☐	☐
I can understand a manual.	☐	☐	☐	☐
I can pick up the main information from machine instructions.	☐	☐	☐	☐
I can communicate with a client about maintenance.	☐	☐	☐	☐
I can write a follow-up letter to a customer.	☐	☐	☐	☐

***Note**: Distinction, Merit, Pass, Fail

New Words and Expressions

Reading A

New Words

amplify /ˈæmpləˌfaɪ/ v. 放大, 增强
board /bɔrd/ n. 板材
eliminate /ɪˈlɪməˌnet/ v. 消除
hardware /ˈhɑrdˌwɛr/ n. 五金器具, (电子仪器的)部件
mill /mɪl/ n. 压榨机, 磨粉机
minimize /ˈmɪnəˌmaɪz/ v. 减少到最低限度
mount /maʊnt/ v. 安装; 安置
nut /nʌt/ n. 螺母, 螺帽
platform /ˈplætˌfɔrm/ n. 平台
pre-finished /ˌpriˈfɪnɪʃt/ a. 抛光好的
screw /skru/ n. 螺丝钉
secure /sɪˈkjʊr/ v. 确保
storage /ˈstɔrɪdʒ/ n. 贮藏
tip /tɪp/ vi. 变倾斜
vibration /vaɪˈbreʃən/ n. 振动
washer /ˈwɑʃɚ/ n. 垫圈
workbench /ˈwɝkbɛntʃ/ n. 工作台, 如机械工人、木匠或珠宝匠的工作台

Phrases & Expressions

act as 担任, 充当
allow for 考虑到, 为了……
be secured to 固定在, 安装在
put away 放好
with washers and nuts 带着垫圈和螺帽

Reading B

New Words

acid /ˈæsɪd/ n. (化) 酸
anti-corrosion /ˌæntɪkəˈroʒən/ n. 防侵蚀
file /faɪl/ n. 锉刀
grease /gris/ n. 润滑油
groove /gruv/ n. 长而窄的槽
multi-lube /ˈmʌltɪˈlub/ n. 多功能润滑油/润滑剂
nail /nel/ n. 指甲, 钉子
paste /pest/ n. 糊状物, 膏
penetrate /ˈpɛnəˌtret/ v. 穿透
polish /ˈpɑlɪʃ/ v. 磨光, 擦亮
rail /rel/ n. 轨道
residue /ˈrɛzɪˌdju/ n. 残余, 滤渣
resin /ˈrɛzn/ n. 树脂
rusty /ˈrʌstɪ/ a. 生锈的, 腐蚀的
solution /səˈluʃən/ n. 溶液
spindle /ˈspɪndl/ n. 主轴
spray /spre/ n. 喷雾
wedge /wɛdʒ/ n. 楔

Phrases & Expressions

be bound to 一定要, 一定会
fit in 相合
on no account 决不
sink in 渗入, 浸透

Technical Terms

guide rail 导轨
oil residue 油渣
spindle thread 螺纹轴

Vocabulary and Structure

Task 1   Write out the words in Reading A according to their meanings in the right column. The first letter of each word has been given.

s_____	to make firm or tight; fasten
m_____	to fix something in position for use
m_____	to reduce to the smallest possible amount, extent, size, or degree
a_____	capable of being gotten; obtainable
e_____	to completely get rid of something that is unnecessary or unwanted
l_____	to direct or carry from a lower to a higher position; raise
a_____	to increase something in size or strength
v_____	vibrating movement or sensation

Task 2   Fill in each blank with the appropriate form of the word given in the brackets.

1. The central aim of the government is to keep social (stable) _____ and economic growth.

2. The (similar) _____ between the two reports suggests that one person wrote both.

3. I was amazed at her (able) _____ to cope with the difficult situation.

4. We make good use of special (plastic) _____ in place of metal.

5. During this stage the atomic system acts as an energy (store) _____ mechanism.

6. There is no point in applying this fertilizer regardless of the (available) _____ of grass.

7. This (arrange) _____ naturally resulted in frequent complaints by the forest inhabitants about illegal extortion by the foresters.

8. The young people spoke at (long) _____ about their deep-sea exploration.

9. Management of human resources is a key to success in any (operate) _____, particularly a family operation.

10. Even at full speed the ship's engines cause very little (vibrate) _____.

Task 3 Complete the following sentences with the words or phrases given below. Change the form if necessary.

| keep from | mount | put away | minimize | in use |
| in case of | on no account | bound to | with the aid of | attach to |

1. The old farmer is walking in the street slowly _____ a cane.
2. He _____ the value of her contribution to his research so that he got all the praise.
3. She's got a few thousand pounds _____ for her retirement.
4. Jane left him folding the coat and began to _____ the stairs.
5. If you undertake the project, you are _____ encounter difficulties.
6. Prof. Smith _____ the medical college as a guest professor.
7. David can't _____ talking about his trip because he is just back from China.
8. We will always come _____ extreme necessity, i.e. if we are badly needed.
9. "If the money is not _____ now," he thought to himself, "I shall not be able to buy bread or wine with it."
10. We must always go into the whys and wherefores of anything. _____ should we follow anyone like sheep (盲从他人).

Task 4 Make sentences with the same pattern as is shown in the example.

Example 1: rubber feet/attach/each corner/the bottom/the mounting boards

— Rubber feet should <u>be attached</u> at each corner on the bottom of the mounting boards.

Example 2: the mill/mount/a similar manner/a pre-finished shelf board/rubber feet/use 10 screws/attach/the mill/the board

— The mill may be mounted in a similar manner on a pre-finished shelf board with rubber feet using 10 screws to <u>attach</u> the mill to the board.

1. he/took out his first patent (专利)/at the age of 24/after/inventing a portable alarm/attach/watches and clocks

   _____

2. fourteen tiny parts/attach/the board/bits of fine wire

   _____

3. the camper's tent/hold/a firm position/ropes/which/attach/stakes

   _____

4. attach/recent photograph/your application form

5. great importance/attach/these problems/and/effective measures/adopt/resolve them

> *Example 1:* Spray plenty of anti-corrosion solution onto the spindles and guide shafts and let the solution sink in/the old grease and oil residue can be eliminated
> — Spray plenty of anti-corrosion solution onto the spindles and guide shafts and let the solution sink in *so that* the old grease and oil residue can be eliminated.
>
> *Example 2:* he got up early/he might catch the first bus
> — He got up early *so that* he might catch the first bus.

1. her parents made many sacrifices/she could go to university

2. he had saved enough money/he could live in comfort

3. he wore a mask/no one would recognize him

4. he saved up his money/he might go abroad for his vacation

5. place the infant upright/he is looking over your shoulder

Task 5 Translate the following sentences into English using the words or phrases given in the brackets.

1. It is necessary to _____ (把机床安装在一块木板上). (mount... to...)

2. Rubber feet _____ (垫在安装板底部的) help minimize the noise and vibration from the motor. (attach to)

3. The machine can _____ (被固定在安装板上) using four screws with washers and nuts. (be secured to)

4. Snack Bar is the place _____ (供应快餐的地方). (available)

5. We'd better start earlier; _____ (我们应考虑到路上交通会有耽搁). (allow for)

Grammar

Non-finite Verbs

Task 1   Complete the sentences with infinitives or gerunds.

1. Dan enjoys *reading* (read) science fiction.
2. Cherry suggested _____ (see) a movie after work.
3. Luckily I managed _____ (find) my way to the hotel.
4. Where did you learn _____ (speak) Spanish? Was it in Spain or in Latin America?
5. Do you mind _____ (help) me translate this letter?
6. He asked _____ (talk) to the store manager.
7. I considered _____ (apply) for the job but in the end I decided against it.
8. If he keeps _____ (come) to work late, he's going to get fired!
9. Debbie plans _____ (study) abroad next year.
10. David failed _____ (pay) his electricity bill.

Task 2   Write "G" in the blank if the underlined part is a gerund (phrase) and "P" if it is a present participle (phrase).

1. Writing an English composition is not easy.   G
2. The story he told us was very interesting.   _____
3. Taking this kind of medicine is useless.   _____
4. I suggested asking his brother for some money.   _____
5. My job is teaching.   _____
6. The play is exciting.   _____
7. I heard the girl singing in the classroom.   _____
8. This is a new washing machine.   _____
9. The man talking with my father is Mr. Wang.   _____
10. Looking from the hill, you can see the whole town.   _____

Task 3 Complete the following conversation with infinitives or gerunds.

Carl: I'm really looking forward to our holiday. I remember 1 _visiting_ (visit) Brazil with you 10 years ago. That was a great trip.

Amy: Yes... Tina, I hope you didn't forget 2 _____ (switch off) the lights in the house!

Tina: Of course I didn't! Stop 3 _____ (worry) about the apartment! Everything's OK. Mom, I hope you remembered 4 _____ (pack) my new toy.

Amy: Wait a minute... I try 5 _____ (think). Mmm, yes, I did. It's in your brown bag. Oh dear... I do things and then I forget 6 _____ (do) them!

Carl: Yes, I still remember 7 _____ (go) to Brazil and you were only interested in talking about carnival. You risked 8 _____ (be trampled) and after the holiday you couldn't remember 9 _____ (do) all this.

Amy: By the way, where are our plane tickets? I hope you didn't forget 10 _____ (bring) them with you!

Task 4 Complete the following conversation with infinitives. Omit "to" if necessary.

A: Are you sure you'll be all right?

B: Honey, sorry to have you 1 _worry_ (worry) about me, but I'm not a child and I can manage 2 _____ (look) after myself.

A: Alright.

B: Some friends have invited me 3 _____ (visit) them. I haven't seen them for a long time.

A: It'll be nice for you to see your old friends again. I just know you're going to have lots of fun. Let me 4 _____ (buy) you a magazine 5 _____ (read) on the train.

B: I can't read when I'm traveling. It makes me 6 _____ (feel) sick, even in a train. I'd rather just 7 _____ (look) out of the window.

A: OK. Well, you'd better 8 _____ (get) on. I think it's about to leave. Oh, did I remind you 9 _____ (change) at York?

B: Yes, you did. Don't worry, I won't forget. I know perfectly well how 10 _____ (get) there.

非谓语动词

不能作谓语的动词称为非谓语动词，包括不定式 (infinitive)、动名词 (gerund) 和分词 (participle)。

1. 不定式与动名词

1) 动名词表示一般或抽象的动作，而不定式则往往表示具体或一次性的动作。
2) 某些动词后只能接不定式作宾语，如：afford, ask, decide, demand, expect, plan 等；而另一些动词后需接动名词作宾语，如：enjoy, finish, avoid, consider, mind, deny, risk 等。
3) 有些动词后既可接不定式也可接动名词，但二者意义有较大的区别，如：forget, stop, remember, regret, try 等。

2. 不定式与分词

1) 很多动词后接不定式作宾语补足语，如：ask, encourage, allow, cause, invite, request 等。
2) 感官动词 feel, see, watch, hear 等和使役动词 have, let, make 后既可接分词也可接不带 to 的不定式作宾语补足语。接现在分词 (present participle) 表示动作正在进行，接过去分词 (past participle) 表示被动，接不定式表示动作已经完成。

→ I heard her singing an English song when I passed by her room yesterday.

→ I heard her sing an English song just now.

→ I heard the English song sung many times.

3) 不定式和分词在句中可以作状语。不定式作状语主要表示目的或结果，分词作状语表示时间、原因、方式、条件等。现在分词与句子主语是主动关系，过去分词与句子主语是被动关系。

→ To hide my emotion, I buried my face in my hands.

→ Not knowing what to do, he went to his parents for help.

→ Given more attention, the trees could have grown better.

3. 动名词和现在分词

1) 动名词和现在分词作表语时，动名词表示主语的具体内容，主语和表语的位置常常可以互换；现在分词表示主语的特点和所处的状态，两者的位置不能互换。

→ My job is teaching English. = Teaching English is my job. (动名词)

→ The film is disappointing. (现在分词)

2) 动名词和现在分词作定语时，动名词表示所修饰词的用途，现在分词所修饰的词与分词在逻辑上构成主谓关系。

→ This is a new washing machine. (动名词)

→ The man talking with my father is Mr. Wang. (现在分词)

Comprehensive Exercises

Task 1 Complete the following dialogs with infinitives or gerunds.

1. **A:** I hear you sometimes sail to France in your boat.
   B: That's right. I really enjoy _____.
2. **A:** Are you going to organize our trip?
   B: Yes, of course. I've agreed _____.
3. **A:** You wear a uniform at work, don't you?
   B: Yes, I have to, although I dislike _____ it.
4. **A:** Do you think they'll approve the plan?
   B: Yes, I'm quite sure they'll decide _____ it.
5. **A:** What time will you be back?
   B: Oh, I expect _____ back sometime around nine.
6. **A:** Did I remind you about the dinner tonight?
   B: Yes, thank you. You keep _____ me.
7. **A:** Do you still work at the post office?
   B: No, I gave up _____ there last year.
8. **A:** Has ICM bought the company?
   B: Well, they've offered _____ it.
9. **A:** I'm sorry you had to wait all that time.
   B: Oh, it's all right. I didn't mind _____.
10. **A:** Do you think your decision was the right one?
    B: Yes, luckily. In the end it proved _____ the best thing for everyone.

Task 2 Rewrite the following sentences after the examples.

> *Example 1:* My father said I could use his car.
> —— My father allowed *me to use his car*.
>
> *Example 2:* Don't stop him from doing what he wants.
> —— Let *him do what he wants*.

1. I think you should know the truth.
   → I want _____.
2. He looks older when he wears glasses.
   → Glasses make _____.
3. My lawyer said I shouldn't say anything to the police.
   → My lawyer advised _____.
4. Remind me to phone my sister tomorrow morning.
   → Don't let _____.
5. I was told that I shouldn't believe anything he says.
   → I was warned _____.

6. The film was very sad and I cried.
→ The sad film made _____.

7. If you have a car, you are able to travel round more easily.
→ Having a car enables _____.

8. The Customs officer made Sally open her case.
→ Sally was made _____.

Task 3 Rewrite each sentence by replacing the underlined part with a participle after the examples.

> *Example 1:* The boy who carried a blue parcel crossed the street.
> —— The boy *carrying a blue parcel* crossed the street.
>
> *Example 2:* When you see it in the air, the mountain looks very small.
> —— *Seen in the air*, the mountain looks very small.

1. He went out and shut the door behind him.
   _____

2. He was a good boy and helped his mother in the kitchen.
   _____

3. I saw the naughty boy. He was hitting the dog.
   _____

4. She had finished her degree and started to work for an international company.
   _____

5. Since she didn't hear the doorbell, she missed the delivery.
   _____

6. The books which were sent to us are for my aunt.
   _____

7. The girl who was picked up by her brother was very nice.
   _____

Task 4 Complete the following passage with the proper form of the words given in the brackets.

Yuri was in his first year at university, studying history. He was rather a lazy student, and tended to avoid 1 _____ (work) whenever he could. In the middle of the semester, his history professor gave an assignment, due in two weeks. Yuri intended 2 _____ (do) the assignment, but he postponed 3 _____ (write) it for a week. The following week, he forgot 4 _____ (do) it. The night before the assignment was due, he suddenly remembered it, and rushed to the library. He tried 5 _____ (read) as much as possible on the topic, but there wasn't enough time. Yuri considered 6 _____ (ask) for more time to do his paper, but the history professor was known to be very tough on students, so finally he decided 7 _____ (cheat) and copy his paper from somewhere else. He found an old article on the same topic, and quickly typed it out. The next day, he submitted the paper. The following week, he was alarmed to see the professor approaching him, looking angry. "Is this your own work, or did you copy it?" asked the professor. Yuri denied 8 _____ (copy) the paper. "If you expect me 9 _____ (believe) that, you must be very stupid," said the professor. "Every word is taken from an article I wrote myself five years ago. Did you really think I would forget 10 _____ (write) it?"

Fun Time

A Humor

A Russian, a Cuban, an American businessman and an American lawyer were on a train traveling across Europe. The Russian took out a large bottle of Vodka, poured each of his companions a drink and hurled the semi-full bottle out of the window.

"Why did you do that?" asked the American businessman.

"Vodka is plentiful in my country," said the Russian. "In fact, we have more than we will ever use."

A little later, the Cuban passed around fine Havana cigars. He took a couple of puffs of his and then tossed it out of the window.

"I thought the Cuban economy was suffering," the businessman said. "Yet you threw that perfectly good cigar away."

"Cigars," the Cuban replied, "are a dime a dozen in Cuba. We have more of them than we know what to do with."

The American businessman sat in silence for a moment, then he got up, grabbed the lawyer and threw him out of the window.

UNIT 7

Operational & Technical Management

Unit Objectives

After studying this unit, you are able to:

- describe the advantages of a PDM used in mold design
- tell the responsibilities of a warehouse worker and a warehouse manager
- talk with other people about a PDM system
- report to the manager on running the warehouse or technology innovation
- draft a mini-report on reform of a company

English for Mechanical & Electrical Engineering

Warming-up

Task 1   As the president of a machine company, you may experience the following situations. Match each situation with its corresponding picture.

☐ Examining the drawings of a new mold development on computer.

☐ Presiding a meeting about innovation of data management.

☐ Inspecting the workshop of the company.

☐ Discussing the warehousing issues with the warehouse worker.

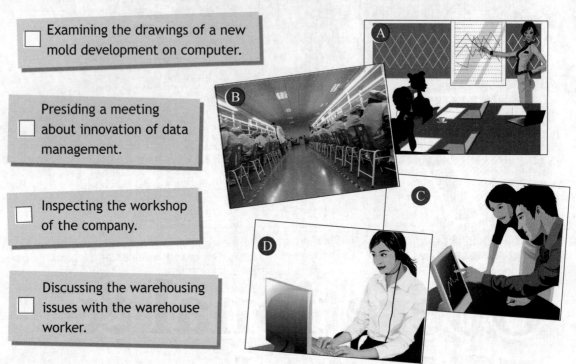

Task 2   The following things are what you usually does for operational and technical management. Arrange them in the order of time and explain.

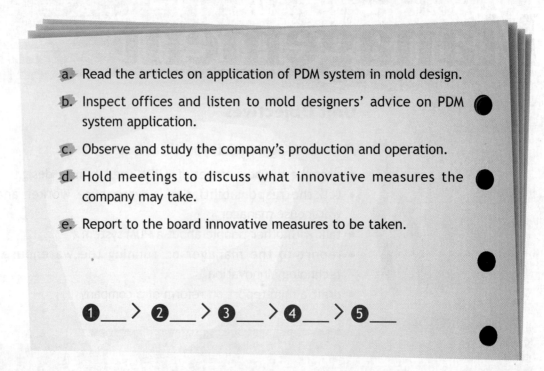

a. Read the articles on application of PDM system in mold design.

b. Inspect offices and listen to mold designers' advice on PDM system application.

c. Observe and study the company's production and operation.

d. Hold meetings to discuss what innovative measures the company may take.

e. Report to the board innovative measures to be taken.

❶___ > ❷___ > ❸___ > ❹___ > ❺___

Reading A

Task 1   Before reading the passage, try to answer the questions about PDM system.

1. Have you ever heard about *PDM*? Do you know its full name?
2. Have you ever heard about any other computer-based system used in mold design?

PDM System and Mold Design

Product Data Management (PDM) system, a computer-based system which electronically maintains the organizational data, promises moldmakers efficient data management and helps them to save time, control costs and ensure quality. The following tips will help moldmakers learn how PDM can meet their data management challenges and understand which PDM system is best for them.

Distinguish between standard and custom parts

In the course of developing molds, moldmakers accumulate a range of standard parts, which they purchase from suppliers, and custom parts, which are unique and require additional work. However, they want to avoid using a custom part when a standard part will do the job.

By using a PDM system, moldmakers can use item numbers to distinguish between standard and custom parts, and use that information to maximize their use of more cost-effective, standardized parts.

Facilitate design reuse, innovation and adaptation

Reusing existing designs and adapting them for new applications is far preferable to starting every mold development project with a blank slate. A PDM system can help mold designers find existing mold designs that are similar to the project at hand, which they can then easily adapt for new jobs, saving time while accelerating estimating and quoting functions.

An integrated PDM system can automatically update BOM (bills of materials) information as mold revisions are made. With an integrated PDM system you can search for and find mold components quickly and easily using a wide range of different properties. In addition to facilitating design reuse, a PDM system encourages innovation because it allows designers to explore what-if scenarios, secure in the knowledge that they can easily roll back to previous versions if a particular idea does not pan out.

Automate workflows

By creating a foundation for more effective design data management, PDM systems can actually enable moldmakers to optimize their development processes through the use of automated workflows. Because PDM systems eliminate any questions or guesswork regarding the validity and currency of design data, moldmakers can create a workflow structure around the data and improve productivity.

In sum, no matter what type of molding a

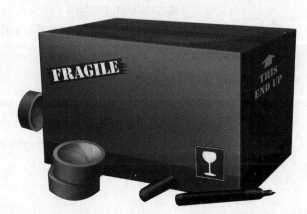

moldmaker does, data management has become a necessity for competing successfully in a global market. With these tips, moldmakers can not only understand how PDM can meet their data management challenges, but also determine which PDM system can help them manage mold design data most effectively.

Task 2　Read the passage. Match each function of a PDM system in the left column with the thing(s) it can achieve in the right column.

automating workflow	arrangement of standard parts and custom parts
distinguishing standard and custom parts	adaptation of original design
facilitating design reuse, innovation and adaptation	creating a workflow structure
	reusing some design data
	updating the bills of materials automatically

Task 3　Read the passage again and decide whether the following statements are true (T) or false (F).

☐ 1. A PDM system can help moldmakers improve the efficiency of data management.
☐ 2. Application of a PDM system in mold design can help the users to ensure the quality of their products.
☐ 3. In developing molds, custom parts are usually unique and often require additional work.
☐ 4. Even if the PDM system is adopted in the mold development, the custom parts and standard parts are still hard to be distinguished.
☐ 5. A new mold development project like a new written book always starts with a blank plate.
☐ 6. A PDM system can help mold designers find existing mold designs that are similar to the project at hand.
☐ 7. An integrated PDM system can facilitate a new mold development.
☐ 8. Using the PDM system in mold development, you can easily obtain some new ideas.
☐ 9. A PDM system can help you to get rid of some doubts and solve some problems in your mold design process.
☐ 10. To adopt a PDM system is one thing, and to improve the productivity is another thing.

Task 4　A PDM system may help mold designers to manage different types of mold design data by integrating different design formats (格式). Discuss with your classmates and list some formats used in the mold design.

Unit 7 Operational & Technical Management

Listening

Task 1 Listen to the conversation and match the people with the correct information.

- an associate manager
- thinks the PDM system is a good thing
- works in the office next to Mr. Stein's
- manager of Mold Design Department
- is worried about application of PDM

Task 2 Mr. Hamlin is talking to Ann Richard, president of Blue-sky Machine Company, about how to save the cost. Listen to the conversation and fill in the blanks with what you hear.

Ann Richard: Good morning, Mr. Hamlin. I'm glad you have brought us some information about mold making techniques in the European countries. Do you have any 1 _____?

Mr. Hamlin: Actually, I just come to report to you some of my ideas about how to 2 _____ cost in the mold making process.

Ann Richard: Well, that is great.

Mr. Hamlin: Mrs. Richard, I think it is necessary for us to 3 _____ some equipment and reform the working procedure.

Ann Richard: Yes, the competition in the mold making industry becomes more and more 4 _____. We have to upgrade our equipment systematically and meanwhile we have to 5 _____ our employees to learn and use new technology.

Mr. Hamlin: To tell the truth, ever since the PDM system was 6 _____ in our mold designing, our working efficiency has been greatly improved. The 7 _____ mold design cost has been greatly reduced.

Ann Richard: Great. Technology is 8 _____; we cannot lag behind the time. As you have suggested, we have also been considering importing some 9 _____ equipment.

Mr. Hamlin: And I think we may employ some graduates from 10 _____ universities.

Ann Richard: Yes, quite right.

Task 3 Helen, a warehouse manager, is talking with her colleague Brad. Listen to the conversation and choose the best answer to each question.

New Words
welding machine   焊接机器
warehouse *n.*   仓库

1. Which company has placed a big order for the welding machines?
   A. A British company.
   B. An American company.
   C. A German company.
   D. A Canadian company.

2. Why does Helen believe the German order has brought the company a great challenge?
   A. Because the company can earn a lot from the order.
   B. Because the company has to employ more workers there.
   C. Because the company needs more technicians to help them.
   D. Because the company has to produce the machine according to the European standards.

3. Why haven't the workers pack the assembled machine?
   A. They have to adjust and test the machines.
   B. They have to ask for permission from the general manager.
   C. They are waiting for the German experts' permission.
   D. They don't know how to pack those machines.

4. When will the first batch of products be moved to the warehouse?
   A. This morning.
   B. Tomorrow morning.
   C. This evening.
   D. Tomorrow evening.

5. What can you infer from the dialog?
   A. It is easy for a Chinese company to meet the European standards.
   B. The warehouse is an important part in the supply chain.
   C. Many Germany companies are now interested in the Chinese products.
   D. It is not too difficult for a company to enter the European market.

Unit 7　Operational & Technical Management

Task 4　Helen is checking the goods that were received yesterday. Listen to the conversation and take notes of the information you hear.

> Arrival time:
> Series Number:
> Quantity:
> Place of storage:

Task 5　Helen is talking with Brad about modern warehouse management. Listen to the conversation and answer the following questions.

New Word

zero inventory　　零库存

1. What role does a warehouse play in a modern business?

2. What is a warehouse worker's job involved with?

3. How can a company realize a zero-inventory strategy?

4. How can a warehouse be modernized?

5. Why can a modernized warehouse help a company save cost?

English for Mechanical & Electrical Engineering

Speaking

Task 1 Work in pairs. Practice making short conversations with the words provided according to the example below.

> Example: Mrs. Richard/Mr. Hamlin/mold designing/PDM system
>
> **A:** Good morning, Mrs. Richard.
> **B:** Good morning, Mr. Hamlin.
> **A:** I'm coming to report to you our reform in mold design.
> **B:** What is it?
> **A:** We try to integrate the PDM system into our original system.
> **B:** Great! I'm glad the new software is used in our company!

1. Mrs. Stein/Mr. White/mold design/3D CAD
2. Ms. Adey/Mr. Hillman/machine manufacturing/some new materials
3. Mrs. Peterson/Mr. Johnston/warehouse management/a new system

Task 2 Work in pairs. Suppose Ann Richard, president of Blue-sky Machine Company wants to get some advice from, Andrew Johnson, manager of the Manufacturing Department. Practice making a conversation. You may use the phrases or expressions listed below.

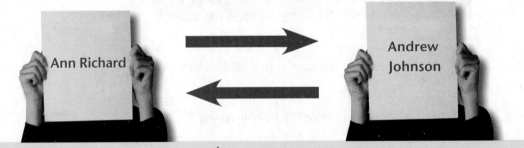

Ann Richard	Andrew Johnson
Sit down, please.	I come here to report...
Have you got any ideas to...	We need to import some advanced machines...
How much should we invest in...	We at least have to invest...
How soon can we get back?	At most in one year.
Any other suggestions?	We need to reform...
Any specific idea?	I've been considering this issue... but...

Unit 7　Operational & Technical Management

Task 3　Work in pairs. Suppose you are Helen, who is talking with her assistant Jane. Make a conversation according to the instructions below.

Helen
Greet.

Ask for some information about the newly installed WMS (warehouse management system) software.

Ask for further suggestions.

Consider the feasibility of the investment.

Jane
Greet.

Stress the advantages of the system.

Suggest applying the automated delivery system.

Suggest making some research.

Task 4　Work in pairs. Helen needs to work closely with the Manufacturing Department. Now Andrew Johnson is asking some information about the inventory of certain materials. Make a conversation with your partner with the help of the instructions below.

Andrew Johnson, Manager of the Manufacturing Department
Identify who you are.

Appreciate the services the Warehouse Department has provided.

Inquire the storage information of certain raw materials.

Express the urgent need to enlarge the workshop.

Helen, Manager of the Warehouse
Greet Andrew Johnson.

Offer thanks for his complimentary comment on your work.

Provide the necessary information inquired.

Ask some advice about the innovation of the warehouse management.

Reading B

Responsibilities of the Warehouse Worker and the Warehouse Manager

Warehouses are where all types of goods and products—both perishable and non-perishable—are stored, ready for distribution. Warehouse workers make sure that stock is stored in the right place, safely, securely and in the correct conditions. They then ensure that the required stock is ready for collection when it is needed.

The Warehouse Worker

The job of a warehouse worker may include:
• checking goods in;
• placing them in their correct location in the warehouse;
• signing delivery notes;
• picking goods from storage and packing them ready for dispatch;
• moving goods around the warehouse by hand and by using lifting gear and fork lift trucks;
• stock keeping—keeping computerized checks of the goods in the warehouse and making sure the numbers in stock match the records;
• reporting any problems, such as faulty or broken goods, to the warehouse manager.

The Warehouse Manager

Warehouse managers are responsible for the efficient running of the warehouse and are in charge of the workforce. Their work can vary from season to season depending on the amount of stock to be moved in and out of the warehouse.

Their job covers:
• planning the movement of goods into and out of the warehouse;
• arranging for the storage of goods and deciding their position in the warehouse;
• keeping track of stock levels;
• making sure goods are stored correctly and safely, following the regulations for the storage of goods such as chemicals, drugs, alcohol and food;
• planning work schedules for staff;
• organizing recruitment and training of staff;
• being responsible for health and safety standards and discipline in the workplace;
• looking after the security of the warehouse;
• scheduling repair and maintenance work, making sure it is carried out properly while causing as little disruption to the running of the warehouse as possible;
• working closely with other departments such as transport and production.

Most warehouses now use computerized stock control systems, so both workers and managers are likely to use computer equipment including handheld systems.

Unit 7 Operational & Technical Management

Task 1 Read the passage and tick the responsibilities of a warehouse manager.

1. checking goods in [   ]
2. planning the movement of goods [   ]
3. following the tracks of stock levels [   ]
4. placing goods in their correct location in the warehouse [   ]
5. looking after the security of the warehouse [   ]
6. signing delivery notes [   ]
7. making sure goods are stored correctly and safely [   ]
8. working out a work plan for staff [   ]
9. moving goods around the warehouse [   ]
10. organizing training of workers [   ]

Task 2 Match the English expressions with their Chinese meanings.

1. non-perishable products
2. check in
3. sign a delivery note
4. ready for dispatch
5. fork lift truck
6. organize a recruitment
7. schedule repair and maintenance work
8. computerized stock control system

a. 组织招聘新人
b. 签署货运单
c. 叉车
d. 安排维修和保养工作日程
e. 准备好发货
f. 不易腐产品
g. 计算机库存管理系统
h. 签收，检验货物入库

Task 3 Translate the following paragraph into Chinese.

Under the management environment of global supply chains, raw materials and goods storage will be under control effectively, but it does not mean that the areas of warehouses will be reduced greatly. On the contrary, areas of single warehouses will become bigger and bigger, while total areas of social warehouses keep growing year by year.

Writing

Task 1 Suppose you are Philip Ball who works in the warehouse. You have received a batch of welding machines from the Manufacturing Department at 8 am on Friday, July 24, 2020. However, you have found that the welding machines DF 140-6 are two sets short. Fill in the necessary information to confirm the goods the warehouse has received and show some discrepancies (差异).

Certificate of Warehousing by the Consignee (产品入库证明)

I/We hereby certify that the consignment arrived at _____ on _____ and that the goods conform in all respects to the description given _____ the following discrepancies, and that they have been _____ under Entry No. 56008 of the register maintained in the warehouse.

Particulars of Discrepancies

No. and description of packages not received.	Quantity short received	Duty payable on the shortage
		USD 7.10

Place: Yongtai Road Warehouse
Date: _____
Signature of consignee(s): _____

Notes:

certificate	证明, 证明书
consignee	受托者, 收件人
hereby	因此, 据此
consignment	发货, 交货

Unit 7 Operational & Technical Management

Task 2 Suppose you are Helen. Write a mini-report to Ann Richard, president of Blue-sky Machine Company, about innovative ideas to reform the warehouse management.

To: Ann Richard
From: Helen, manager of Warehouse Department
Date: July 28, 2020
Subject: Reforming the warehouse management

In this report I will present some innovative ideas to reform our warehouse management.

The fast development of the Manufacturing Department urges us to enlarge the capacity of the old warehouse and improve the efficiency of management. Therefore, we suggest that some measures be taken.

Firstly, _____
_____
_____
_____
_____

Secondly, _____
_____
_____
_____
_____

Thirdly, _____
_____
_____
_____
_____.

Finally, I recommend that we improve our delivery system to reduce the time of loading and unloading goods from and to the warehouse.

Project

Project Guidelines

This project aims to go through the process of a product development and improvement in a company. The whole process is divided into three steps. Step One is searching for information on the welding machine. Step Two focuses on technological challenges and customers' needs. Step Three rests on the discussion about product development and improvement.

Please follow the *Task Description* to complete the project.

Task Description

 Step One

- Set up a small group of 4-6 students in your class;
- Search online for information on welding machines;
- Share your information from the Internet or other resources with your group members.

 Step Two

- Present the strengths of your group in the welding machine development with the help of information from the Internet or other resources;
- Show what innovation your group will carry out and explain;
- Divide your group into two parts: one side being the group member of the Technology Department in a company, the other being the potential customers;
- Take turns to play each role: the customer group showing what kind of welding machine they need, the design group promising what innovative measures they will take.

 Step Three

- Both the developing group and customer group note down the requirements and promises of the two sides;
- Make a summary about how a new product is developed in a company and present it to the class.

Self-evaluation

Rate your progress in this unit.	D	M	P	F*
I can describe the advantages of a PDM used in mold design.	☐	☐	☐	☐
I can tell the responsibilities of a warehouse worker and a warehouse manager.	☐	☐	☐	☐
I can talk with other people about a PDM system.	☐	☐	☐	☐
I can report to the manager on running the warehouse or technology innovation.	☐	☐	☐	☐
I can fill in the forms related to goods in the warehouse.	☐	☐	☐	☐
I can draft a mini-report on reform of a company.	☐	☐	☐	☐

* *Note*: Distinction, Merit, Pass, Fail

New Words and Expressions

Reading A

New Words

accumulate /əˈkjumjuˌlet/ v. 积累，积聚
adaptation /ˌædæpˈteʃən/ n. 适应，改编
additional /əˈdɪʃənl/ a. 另外的，附加的
automate /ˈɔtəˌmet/ v. (使)自动化
cost-effective /ˈkɔstɪˌfɛktɪv/ a. 有成本效益的，划算的
currency /ˈkɚənsi/ n. 流通
distinguish /dɪˈstɪŋgwɪʃ/ v. 区别，使显出不同特征
facilitate /fəˈsɪləˌtet/ v. 推动，促进
foundation /faunˈdeʃən/ n. 基础，根本
innovation /ˌɪnəˈveʃən/ n. 革新，创新
maximize /ˈmæksəˌmaɪz/ v. 最佳化，最大化
optimize /ˈɑptəˌmaɪz/ v. 使最优化
preferable /ˈprɛfrəbl/ a. 更可取的，更好的
property /ˈprɑpətɪ/ n. 属性，特性
quote /kwot/ v. 报价
scenario /sɪˈnærɪˌo/ n. 预料或期望的一系列事件模式
slate /slet/ n. 石板，石片
update /ʌpˈdet/ v. 更新
validity /vəˈlɪdətɪ/ n. 有效性，合法性
version /ˈvɚʒən/ n. 版本
what-if /ˌwɑtɪf/ n. 假定分析
workflow /ˈwɚkflo/ n. 工作流程

Phrases & Expressions

custom parts 客户定制/特需的部件
integrated PDM system 集成PDM系统
pan out 结果好，成功
roll back 回转

Reading B

New Words

chemical /ˈkɛmɪkl/ n. 化学制品 a. 化学的
discipline /ˈdɪsəplɪn/ n. 纪律
dispatch /dɪsˈpætʃ/ n. 分派，派遣
disruption /dɪsˈrʌpʃən/ n. 中断，破坏
drug /ˈdrʌg/ n. 药，药物
handheld /ˈhændˌhɛld/ a. 手持式的，便携的
perishable /ˈpɛrɪʃəbl/ a. 容易腐烂的
recruitment /rɪˈkrutmənt/ n. 招募，招聘
regulation /ˌrɛgjəˈleʃən/ n. 规章，规则
staff /stæf/ n. (全体)职员，工作人员
stock /stɑk/ n. 库存
vary /ˈvɛrɪ/ v. 变化，不同

Phrases & Expressions

be likely to 很有可能，倾向于
check... in 核对无误
in stock 库存，现货
keep track of 记录，跟踪

Technical Terms

lifting gear 吊运装置

Vocabulary and Structure

Task 1 Write out the words in Reading A or Reading B according to their meanings in the right column. The first letters are already given.

m_____   to keep something in existence at the same level, standard, etc.
d_____   to recognize and understand the difference between two objects or matters
p_____   to obtain in exchange for money or its equivalent; buy
f_____   to make easy or easier
a_____   the process of changing something to make it suitable for a new situation
a_____   to happen faster than usual or sooner than you expect
a_____   to convert to automatic operation
i_____   the act of introducing something new

Task 2 Fill in each blank with the appropriate form of the word given in the brackets.

1. As a youth he showed no (promise) _____ of becoming a great pianist.

2. The capitalists (资本家) only care about the (accumulate) _____ of wealth.

3. All these islands are very beautiful, and (distinguish) _____ by various qualities.

4. The production of Tibetan medicine has been put on a (standard) _____, normalized and scientific administration track.

5. Peter, my English teacher, never fails to (courage) _____ us to study hard.

6. Clearly, some of the arguments have greater (valid) _____ for some Third World countries than others.

7. Probably the most important component is the (deliver) _____ and evaluation of current information.

8. Put them in a nice, insured, air-conditioned (store) _____ warehouse.

9. It's our duty and (responsible) _____ to protect this pure land—the Antarctic.

10. If family (disrupt) _____ occurs, fathers need to be responsible for economic, social, and psychological support of their children.

Task 3 Complete the following sentences with the words or phrases given below. Change the form if necessary.

| check in | sign | in stock | keep track of | be responsible for |
| update | be likely to | distinguish | preferable to | roll back |

1. Everyone must _____ every morning before going up to their offices.
2. These services _____ be available to us all before long.
3. We have many patterns _____ for you to choose from.
4. How do you _____ these three parts from one another?
5. Make sure you read all the small print before _____.
6. He denied that he _____ the accident.
7. I think riding a bicycle is _____ walking.
8. The pictures _____ the days of childhood to me.
9. Her mother used to _____ every penny she spent.
10. They made out the plans to _____ manufacturing procedures.

Task 4 Make sentences with the same pattern as is shown in the examples.

Example 1: Reusing existing designs/adapting them for new applications/prefer/start every mold development project/with a blank slate
— Reusing existing designs and adapting them for new applications is far *preferable to* starting every mold development project with a blank slate.

Example 2: it/prefer/have her continue staying/in the hospital/for further treatment
— It is *preferable to* have her continue staying in the hospital for further treatment.

1. anything/prefer/have them with us/for the whole week

2. which cities/prefer/for tourists to visit/on their first trip to China

3. writing a paper/prefer/sit in for an examination

4. I still think it/prefer/put off/the establishment of diplomatic relations with that country/for some years

5. anything/prefer/be a helpless witness/to such pain

> *Example 1:* They didn't want to admit to any shortages of food./It might reveal a weakness to be exploited by their enemies.
> — They didn't want to admit to any shortages of food *in case* it might reveal a weakness to be exploited by their enemies.
>
> *Example 2:* British officers were forbidden to keep diaries./They were captured and their secrets betrayed.
> — British officers were forbidden to keep diaries *in case* they were captured and their secrets betrayed.

1. He wasn't going to use his car./Somebody recognized it.

2. I think I ought to stay./Jim suddenly comes back.

3. He refused to say anything./He would say something foolish.

4. I said that I was 10 years younger./I would be turned down for the job.

5. I was afraid to open the door./He would follow me.

Task 5 Translate the following sentences into English using the words or phrases given in the brackets.

1. _____ (通过使用PDM系统) moldmakers can use item numbers to distinguish between standard and custom parts. (by)

2. They can easily roll back to previous versions _____ (如果某一个新主意行不通). (pan out)

3. Their work can vary from season to season, _____ (取决于搬进、运出仓库的货物量). (depend on)

4. Warehouse managers are responsible for the efficient running of the warehouse and _____ (负责管理那里的员工). (in charge of)

5. Warehouses are where all types of goods and products— _____ (既有易腐烂的, 也有不易腐烂的) — are stored, ready for distribution. (both... and)

Grammar

Subjunctive Mood

Task 1 Cross out the incorrect one of the two choices underlined in each sentence.

1. If I miss/~~I'll miss~~ the bus tomorrow morning, I'll get a taxi instead.
2. We'll have to go without John if he doesn't arrive/won't arrive soon.
3. If you don't apologize/won't apologize now, I will never speak to you again.
4. You will have to/have to ring the bank if you renew your credit card next Sunday.
5. If I make some coffee, do you cut/will you cut the cake?
6. If I knew her number, I could ring/can ring her up.
7. Did you work/Would you work harder if you were better paid?
8. If I am/I were you, I'd ask a lawyer for some advice.
9. What will you do/would you do if you were bitten by a snake?
10. They will be offended/would be offended if we didn't accept their invitation.

Task 2 Complete the following conversation between a husband and a wife with the proper form of the verbs in the brackets.

H: I haven't forgotten our wedding anniversary, you know. If you like, I 1 *will book* (book) a table for Wednesday at our favorite restaurant.

W: But our anniversary is on Tuesday. You're playing football then, aren't you? If you cared me, you 2 _____ (not play) football that day.

H: What's the difference? If we 3 _____ (go) out on Wednesday, it'll be just the same. If I 4 _____ (not play), I'd be letting the team down.

W: Yes, I suppose it 5 _____ (be) a disaster if you missed one game. Well, if you 6 _____ (think) more of your friends than you do of me, you can forget the whole thing.

H: I just don't understand you sometimes.

W: If you 7 _____ (think) about it, you'd understand. And I think it 8 _____ (be) better if we forgot all about our wedding anniversary.

H: Don't be silly, honey. If you get into one of your bad moods, it 9 _____ (not do) any good.

W: If you were interested in my feelings, I 10 _____ (not get) into a bad mood.

Task 3 Complete the following conversation between two football fans with the proper form of the verbs in the brackets.

   A: Our team didn't play very well today.
   B: We were awful. But if Billy had taken that easy shot, we would have won.
   A: We didn't deserve to win. It 1 *would have been* (be) pretty unfair if our opponents 2 _____ (lose).
   B: Billy was dreadful. My grandmother 3 _____ (score) if she 4 _____ (be) in that position.
   A: And if Rick 5 _____ (not be) asleep, he 6 _____ (not give) a goal away.
   B: Also, if James 7 _____ (not be) injured when we needed him most, it 8 _____ (be) different.
   B: Yes, we 9 _____ (beat) them if he 10 _____ (be) in good shape.

Task 4 Fill in the blanks with the proper form of the verbs in the brackets.

   A: Our English teacher is too strict with us.
   B: Strict? In what ways?
   A: Well, he insists that we 1 *(should) be* (be) punctual for class.
   B: Yes, every student must be punctual for class.
   A: He demands that we 2 _____ (speak) no Chinese in class.
   B: It is essential that you 3 _____ (have) more opportunities to practice oral English.
   A: He also insists that we 4 _____ (keep) a diary in English.
   B: Well, let's forget your teacher for a while and talk about our plan for the winter vacation.
   A: OK. Do you have any ideas now?
   B: My suggestion is that we 5 _____ (go) to the north part of China.
   A: Great! I have never been there.
   B: But it is very cold there and my father suggests that we 6 _____ (take) enough clothes.

虚拟语气

语气是一种动词形式,表示说话者的意图和态度。英语中的语气有三种:1) 陈述语气;2) 祈使语气;3) 虚拟语气。

虚拟语气表示动作或状态不是客观存在的事实,而是说话人的假设、推测或主观愿望等。

1. 条件句与虚拟语气

	从句谓语动词	主句谓语动词
与现在相反	一般过去时/were	would/could/should/might do
与过去相反	过去完成时	would/could/should/might have done
与将来相反	were to/should do	would/could/should/might do

注:If he comes, he will bring his violin. 此句中的条件句为真实条件句,从句谓语动词用一般现在时,主句谓语动词用 shall/will + 动词原形。

2. should与虚拟语气

在表示命令、建议、请求、紧迫等意义的词后面的名词性从句中,谓语动词要用should + 动词原形,其中should可省略。

It is suggested that we (should) hold a meeting next week. (主语从句)

The guard at the gate insisted that everybody (should) obey the rules. (宾语从句)

My advice is that we (should) get more people to attend the conference. (表语从句)

I made a proposal that we (should) hold a meeting next week. (同位语从句)

3. wish与虚拟语气

wish 后宾语从句的谓语动词可用一般过去时、过去完成时和would/could + 动词原形,分别表示所希望的情况与现在事实相反、与过去事实相反和将来不太可能实现的愿望。

I wish I were as tall as you. (与现在事实相反)

I wish I hadn't said that. (与过去事实相反)

I wish it would rain tomorrow. (将来不太可能实现的愿望)

Comprehensive Exercises

Task 1 Rewrite the following sentences after the example.

Example: It might be fine tomorrow. We'll go for a picnic.
—— *If it is fine tomorrow, we'll go for a picnic.*

1. Nick may arrive a bit early. He can help Peter to get things ready.
2. You might forget to phone. They will have to go without you.
3. He may leave early tomorrow morning. You can ask him to give you a lift.
4. I might have time tonight. I'll finish the novel I'm reading.
5. Tom may still be ill tomorrow. He ought to stay at home.
6. You may finish your work early. You could come for a drink with us.
7. There might be too much work to do. You can ask someone to help him.
8. The office may be closed. In that case, Mark won't be able to get in.

Task 2 Rewrite the following sentences after the example.

Example: I don't have a spare ticket. I can't take you to the concert.
—— *If I had a spare ticket, I would take you to the concert.*

1. John ate too much birthday cake, so he was sick.
2. She drinks too much coffee. She doesn't feel calm.
3. He can't type. He isn't able to operate a computer.
4. They don't understand the problem. They won't find a solution.
5. She is not in your position. She isn't able to advise you.
6. We came home from our holiday early because we ran out of money.
7. I had an accident because I wasn't watching the road.
8. My father didn't earn much money, so life wasn't easy for us.

Task 3 Make sentences beginning with "I wish..." after the example.

Example: The phone in the office has been ringing for five minutes (and you can't answer it).
—— *I wish somebody would answer the phone.*

1. You don't know anything about cars (and your car has just broken down).
   I wish _____.
2. You've eaten too much and now you feel sick.
   I wish _____.
3. You're looking for a job—so far without success. Nobody will give you a job.
   I wish somebody _____.
4. A lot of people drop litter in the street. You don't like this.
   I wish people _____.
5. You're not lying on a beautiful sunny beach (and that's a pity).
   I wish I _____.
6. You live in a big city (and you don't like it).
   I wish I _____.
7. When you were younger, you didn't learn to play a musical instrument. Now you regret it.
   I wish I _____.
8. You've painted the gate red. Now you think that it doesn't look very nice.
   I wish I _____.
9. You want to take some photographs of the park, but you didn't bring your camera.
   I wish I _____.

Task 4 Fill in the blanks with the appropriate form of the verbs given in the brackets.

An old man recently celebrated his 112th birthday and reporters visited him in his mountain village to discover the secret of a long life. "The secret of a long life," he said, "is happiness. If you are happy, you will live a long time." "Are you married?" a reporter asked. "Yes. I married my third wife when I was 102. If you are happily married, you 1 _____ (live) for ever. If it hadn't been for my third wife, I 2 _____ (die) years ago." "What about smoking and drinking?" a reporter asked. "Don't smoke at all and you 3 _____ (feel) well," he said. "Drink two glasses of wine a day and you 4 _____ (be) healthy and happy." "If you could live your life again, what 5 _____ (you do)?" a reporter asked. "I wouldn't change anything. Except, if I had had more sense, I 6 _____ (eat) more yogurt," he chuckled. "Suppose you 7 _____ (can change) one thing in your life, what 8 _____ (you change)?" another reporter asked. "Not much," he replied. "If I 9 _____ (know) I was going to live so long, I 10 _____ (look after) myself better."

Fun Time

Scrap Yard (废料场)

A company had a vast scrap yard in the middle of a desert. Management said, "Someone might steal from it at night." So they created a night watchman (巡夜人, 看守人) position and hired a person for the job.

Then management said, "How does the watchman do his job without instructions?" So they created a Planning Department and hired two people: one person to write the instructions and one person to do time studies.

Then management said, "How will we know the night watchman is doing his tasks correctly?" So they created a Quality Control Department and hired two people: one to do the studies and one to write the reports.

Then management said, "How are these people going to get paid?" So they created the following positions, a timekeeper and a payroll (薪水册) officer; then hired two more people.

Then management said, "Who will be accountable (应负责的) for all of these people?" So they created an administrative (管理的) section and hired three people: an Administrative Officer, an Assistant Administrative Officer, and a Legal Secretary.

Then management said, "We've had this command in operation for one year now and we're $18,000 over budget (超过预算). We have to cut back on overall costs."

So they laid off (解雇) the night watchman.

UNIT 8

Marketing and After-sales Service

Unit Objectives

After studying this unit, you are able to:
- work out marketing strategies for a company
- introduce products to customers
- negotiate mode of payment and delivery with customers
- deal with customers' complaints
- make a survey on customers' feedback to the products and services

Warming-up

Task 1 The following are the daily routine work of a salesman. Arrange them in the order of time and explain.

1. Get customers' feedback to your company.

2. Introduce your products to customers at a trade show.

3. Handle customers' complaints about your products.

4. Discuss details of business with customers.

5. Decide on marketing strategies for your company.

Task 2 Tick the qualities below that you think are most important for a great salesman and explain.

- ○ hard work
- ○ self-confidence
- ○ insight
- ○ leadership
- ○ brazenfaced
- ○ sense of responsibility
- ○ being astute (精明的)
- ○ good luck
- ○ health
- ○ friendliness
- ○ team spirit
- ○ being efficient use of time
- ○ strong will
- ○ knowledgeable
- ○ a large social circle
- ○ ability to handle crisis and cope with difficulties
- ○ ability to communicate effectively

Reading A

Task 1 Before reading the passage, try to answer the questions about niche marketing and after-sales service.

1. Which company do you know has good marketing strategies? Can you give a marketing case?
2. Why should a company offer after-sales service?

Doosan Infracore's Marketing Strategy in China

Doosan Infracore, a manufacturer of construction equipment specializing in excavators and wheel loaders, is making inroads into the Chinese machine tool market. Since its first entry into the Chinese market in 1996, Doosan Infracore has made remarkable achievements based on its localization strategy, niche demand creation, technological innovation and nationwide sales and after-sales service network.

To begin with, Doosan Infracore has introduced a wide range of China-oriented models in Chinese market to meet the needs of local consumers. Doosan Infracore studies China's natural environment and then develops machine tools that best suit Chinese topographical conditions. While a handful of foreign manufacturers have been handling the high-end machine tools in China, Doosan Infracore specializes in low-end models to satisfy China's increasing demand for ordinary machine tools.

Secondly, Doosan Infracore has been staging its in-house machine tools fair in China ever since 2005 to cultivate new customers, expand the order base and support dealers' sales efforts. By

inviting major customers, industry association representatives and government officials to attend its trade shows, Doosan Infracore takes advantage of the shows to demonstrate its latest technologies and production processes and inspire greater trust in Doosan Infracore products.

Thirdly, technological innovation has been central to Doosan Infracore's enhancement in its competitiveness. Doosan Infracore establishes presence in China by making new

technologically advanced products and reinforced its position by following an aggressive network expansion strategy. DH Series excavators are a proof of it. With remarkably increased cab comfort, better visibility, reduced whole body vibrations and noise levels, DH Series excavators appeal to Chinese customers at their release. In addition to excellence in workability, durability, operator convenience and fuel efficiency, the prices are similar to those offered by other Chinese manufacturers, meaning that they are competitive both in price and product quality.

Fourthly, Doosan Infracore has established a nationwide sales and after-sales service network in China in an effort to set up a different image. At present, the network has a total of 280 sales and after-sales service points, the largest organization of its kind in China. The company operates a 24-hour after-sales service system, offering an 8,000-hour free after-sales service guarantee. The company has also made efforts to distinguish itself from others in the Chinese market with prompt decision-making and operational transparency and adopted the market-sensing system intended to respond to the market's demands promptly.

Task 2 The following sentences are the topic sentences of the paragraphs. Read the passage and fill in the blanks.

Paragraph 1: Doosan Infracore is 1 _____ the Chinese machine tool market.

Paragraph 2: Doosan Infracore has introduced 2 _____ in Chinese market to meet the needs of local consumers.

Paragraph 3: Doosan Infracore has been 3 _____ in China ever since 2005.

Paragraph 4: 4 _____ has been central to Doosan Infracore's enhancement in its competitiveness.

Paragraph 5: Doosan Infracore has established 5 _____ in China.

Unit 8 Marketing and After-sales Service

Task 3 Read the passage again and choose the best answer to each question.

1. Doosan Infracore's success in China's market can be attributed to the following marketing strategies except _____.
   A. the localization strategy
   B. a wide variety of distribution channels
   C. niche demand creation
   D. technological innovation

2. Doosan Infracore specializes in low-end models in China because _____.
   A. many foreign manufacturers have been handling the high-end machine tools in China
   B. low-end models need less money
   C. these are the only models they can manufacture
   D. there is an increasing demand for low-end models in China

3. Doosan Infracore stages in-house machine tools fair in China in order to _____.
   A. cultivate new customers
   B. enlarge the order base
   C. support dealers' sales activities
   D. all of the above

4. Which of the following is NOT the reason for DH Series' being accepted by Chinese customers?
   A. Increased cabin comfort.
   B. Fuel economy.
   C. Decreased whole body vibrations and noise levels.
   D. Medium price.

5. Which of the following is NOT true about Doosan Infracore's sales and after-sales service network?
   A. The network helps Doosan Infracore to set up a different image.
   B. The network is the largest organization of its kind in China.
   C. The after-sales service is only available in the daytime.
   D. The customers can enjoy an 8,000-hour free after-sales service.

Task 4 Discuss with your classmates about Doosan Infracore's marketing strategies mentioned in the passage. Can you contribute one more marketing strategy to the company?

Listening

Task 1 Listen to the conversation and match the companies with the correct information.

a. a manufacturer of construction equipment
b. in business line for more than 15 years
c. has three branches in China
d. specializes in excavators and wheel-loaders
e. take part in several infrastructure investment projects in west China
f. its products enjoy wide acceptance in the world market.

Task 2 John Smith talks with Wang Peng, the saleswoman of Doosan Infracore at the trade show. Listen to the conversation and fill in the blanks with what you hear.

Wang Peng: Good morning. May I help you?

John Smith: I 1 _____ if you can give me more information about this excavator model you're showing.

Wang Peng: I'd be glad to help. Would you like a packet of our 2 _____ literature?

John Smith: Thank you. I see this excavator has more 3 _____ operating environment.

Wang Peng: Yes, the specially designed operating 4 _____ provides operator the most space in its class, and its 5 _____ 6 _____ allows easy entrance and exit.

John Smith: Sounds great. Does it perform efficiently?

Wang Peng: Well, this model has a good 7 _____ for its excellence in workability, 8 _____ and fuel efficiency. Its powerful motor and strong front section greatly reduce the body vibrations and noise levels.

John Smith: Remarkable!

Wang Peng: Would you like to tour our company?

John Smith: Yes, if it wouldn't take too long to arrange. I'm 9 _____ 10 _____ fly back to Guangzhou on Wednesday.

Wang Peng: I'm sure we can arrange it before you leave.

John Smith: Thank you for your help.

Wang Peng: You are welcome.

Unit 8　Marketing and After-sales Service

Task 3 John Smith goes to Doosan Infracore to get more information about the excavators. Marketing Manager Alicia Green receives him. Listen to the conversation and tick the correct sentence.

1.
- [ ] John Smith wants to drink coffee.
- [ ] John Smith wants to drink Cola.

2.
- [ ] John Smith is interested in the DX 190W excavators.
- [ ] John Smith is interested in the DX 119W excavators.

3.
- [ ] In John Smith's opinion, the price Alicia Green offers is 10% higher than that of the same products of other brands but acceptable.
- [ ] In John Smith's opinion, the price Alicia Green offers is 10% higher than that of the same products of other brands and unacceptable.

4.
- [ ] Alicia Green offers such a price because his company provides a better service.
- [ ] Alicia Green offers such a price because their products are of better quality.

5.
- [ ] Alicia Green agrees to give John Smith a 2% discount for the size of his order is large.
- [ ] Alicia Green agrees to give John Smith a 2% discount for he is eager to establish a long-term business relationship with John Smith.

Task 4 John Smith calls Alicia Green for more details of the business. Listen to the conversation and answer the following questions.

> **New Words**
> receipt *n.*　　　　　　收到，接到
> covering letters of credit　信用证

1. Why does John Smith call Alicia Green?

   _____

2. When does Alicia Green's company usually deliver the goods after the receipt of the covering letters of credit?

   _____

3. Why does John Smith propose prompt shipment?

   _____

4. Does Alicia Green agree on John Smith's proposal?

   _____

5. What is the mode of payment in this business?

   _____

Task 5 John Smith comes to Doosan Infracore to make a complaint. Cindy Pan, manager of After-sales Department, receives him. Listen to the conversation and complete the sentences with the information you hear.

New Words
bucket *n.*   (挖土机等的) 铲斗, 勺斗
stick cylinder   斗杆油缸

Cindy Pan: Can I help you, sir?
John Smith: Yes. We bought 100 DX 190W excavators from your company 8 months ago and now one of them simply doesn't work.
Cindy Pan: Doesn't work? That's very strange. What's the problem?
John Smith: Poor quality, I guess. When our operator tried to drop the bucket, 1 _____ _____ that the stick cylinder broke into two pieces.
Cindy Pan: I'm sorry to hear that. But technically speaking, 2 _____. They are well accepted for their high efficiency and durability.
John Smith: I'm afraid not. Its alarm system doesn't function properly, either. When the operator uses the excavator to do lifting duties, 3 _____, which makes it impossible to carry out a task.
Cindy Pan: It's unbelievable. Do you have your operator properly trained?
John Smith: Why? Of course. I don't think it's the operator's mistake. Can you replace it with another one?
Cindy Pan: I'm afraid not. 4 _____.
John Smith: Um… Well, actually, we'd like you to exchange it for another one.
Cindy Pan: I'm awfully sorry, sir. We can't do that. The only thing we can do for you now is to send a technician to your company.
John Smith: This is unreasonable. You promised good service when selling excavators to us.
Cindy Pan: Yes, 5 _____ and have the machine fixed for you.
John Smith: All right, then.

Unit 8 Marketing and After-sales Service

Speaking

Task 1 Work in pairs. Practice making short conversations with the words provided according to the example below.

> Example: John Smith/Spark Infrastructure Company/diggers/enter into business relations
>
> A: Good morning. My name is John Smith and I work for Spark Infrastructure Company.
> B: Good morning. How can I help you?
> A: I'm interested in the diggers manufactured by your company. I'm wondering whether it's possible for us to enter into business relations with you.
> B: Yes. Do you mind telling me more about your company?

1. Reyn Peterson/Nanjing Machine Tool Company/wheel loaders/supply goods at once from stock
2. Terry Chris/Shandong Precision Machinery Co., Ltd./skid steer loaders/types of the products
3. Timber Clair/Hebei Measuring and Cutting Tool Company/milling machines/visit the showroom (展室)

Task 2 Work in pairs. Wang Peng is introducing a new model of wheel loader to John Smith. Practice making a conversation. You may use the phrases or expressions listed below.

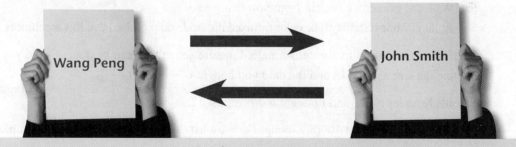

Welcome to our showroom.	I'm interested in...
Spacious cab	Could you show me your catalog?
Compact central monitor panel (中央监控器)	Can you tell me more about...
	How about its operating environment?
Adjustable steering column	Does it perform efficiently?
Low fuel consumption	Thank you very much for the information.
It's nice to talk to you.	

English for Mechanical & Electrical Engineering

Task 3 Work in pairs. Wang Peng is discussing price, terms of payment and delivery with John Smith. Make a conversation according to the instructions below.

Wang Peng

Greet John Smith.

Make a proposal.

Make a more reasonable proposal.

Express thanks and enjoyment.

John Smith

Inquire the price and terms of payment and delivery.

Express polite disagreement.

Express agreement and thanks.

Task 4 Work in pairs. John Smith is making a complaint to Cindy Pan about the wheel loader he just bought. He complains that the central monitor panel gets rusty and the engine makes strange noise when it's moving. Make a conversation with your partner with the help of the sample.

Sample:

A: Hello, sir. What can I do for you?

B: I am Mr. Lee from Anderson & Lewis. I have a complaint to make.

A: I see. I am listening and hope I can be of some help.

B: Well, the printer we bought from you two weeks ago couldn't work yesterday. The color is mixed up. And you promised that this machine is the best.

A: I'm sorry; that's all very unfortunate. I'm sure something can be done. Could you please tell me the size and model and the date you bought it?

B: It's NovaJet 1000i, and I bought it on August 12.

A: OK, that's NovalJet 1000i… bought on August 12… and the color is mixed up… and your name please, sir?

B: Lee Cruise from Anderson & Lewis.

A: Lee Cruise… Anderson & Lewis. OK, sir, I have noted down everything and I am so sorry about the inconvenience you have had. Could you please leave the machine here and we'll send a technician to have it checked and repaired first. What about I call you as soon as we have solutions?

B: No. I hope you understand that a simple repair is just not enough. I demand that you exchange it for another one.

A: If there's really something wrong with the quality of our printer, I will replace it for you.

B: OK, but I hope you can handle it as soon as possible.

Reading B

Invitation to India Machine Tools Show

Dear Sir,

We are pleased to announce the most awaited industrial exhibition, India Machine Tools Show 2009 (3rd IMTOS).

Approved by: ITPO (India Trade Promotion Organization)

Date: July 26–28, 2009

Venue: Pragati Maidan, New Delhi, India

IMTOS 2009, Asia's 2nd largest international industrial exhibition will have on display engineering, machinery, machine tools, automation, material handling equipments, hydraulics & other industrial products and technology. This will be the 3rd IMTOS industrial exhibition that remains to be the most awaited event in the engineering field. Its primary objective is to promote and encourage well-organized expansion and development of the machine tools industry.

It is a matter of pride for us that such an event has been running successfully in India over the past decade. The 3rd IMTOS will have over 1,000 participants, both national and international are expected. We hope to surpass the record of last IMTOS 175,000 visitors, in 3rd IMTOS.

Other than being a successful industrial exhibition IMTOS is also a medium, which can help achieve communication objectives of participants. It holds a prime position in the minds of people; the audience is respectively large in number and is specific to the industrial world. We encourage greater participation from the industry to take a more proactive role, primarily by sponsoring various spatial positions, which can highlight your corporate logo and convey your message to your advantage at the least cost.

For any further information or booth reservation required by you, please do not hesitate to contact us. We will only be happy to respond at the earliest.

We welcome your visit to IMTOS 2009 and wish your ventures the best of success.

Yours sincerely,

Barindrasinh Jhala

Business Development Department of K & D Communication Ltd.

English for Mechanical & Electrical Engineering

Task 1　Read the passage and decide whether the following statements are true (T) or false (F).

☐ 1. IMTOS 2009, Asia's 2nd largest International Industrial Exhibition, is expected to see record-breaking numbers both in terms of exhibitors and visitors.

☐ 2. IMTOS 2009 is a platform for participants from India and the surrounding countries to achieve communication and expansion of their businesses.

☐ 3. The organizer of IMTOS 2009 prefers machine tools to material handling equipments and hydraulics products because the exhibition is a machine tools show.

☐ 4. All the participants' logos will appear in outdoor advertising at the least cost.

☐ 5. Participants have to contact the organizer if they want a booth reservation form, and they are likely to get a prompt reply.

Task 2　Match the English expressions with their Chinese meanings.

1. on display
2. material handling equipments
3. the most awaited event
4. achieve communication objectives
5. hold a prime position
6. take a more proactive role
7. highlight one's corporate logo
8. to one's advantage

a. 实现沟通目标
b. 起更积极的作用
c. 对……有利/有帮助
d. 占重要地位
e. 材料处理设备
f. 展示，展出
g. 最期待的事情
h. 突出企业标识

Task 3　Translate the following paragraph into Chinese.

Please visit CIMT2009, booth W3, B-101, and see how our advanced technology increases the productivity while cutting costs. You'll find everything from the most basic, most economical designs to advanced machines capable of producing complex geometries in a single setup. Visit the booth to explore the innovation, and the attention to service, quality and performance. We'll show you our never-ending efforts to be the world top machine tools maker.

_____
_____
_____
_____

Writing

Task 1 Suppose you are Alicia Green, manager of the Marketing Department in Doosan Infracore. Write a letter of reply to the invitation to IMTOS 2009. Your letter should cover the following points:

1. thanks for the invitation;
2. models you are to display on the trade show;
3. the person your company assigned to make pre-show preparations.

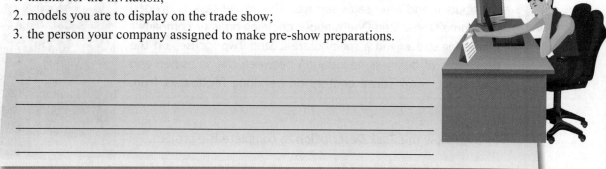

Task 2 Wang Peng took part in IMTOS 2009 in India last month. Suppose you are Wang Peng and fill in the feedback form offered by the organizer of the exhibition.

General Information	
Company	
Country	
Telephone	
Booth Traffic	
○	Good: Steady stream of visitors.
○	Fair: Intermittent (间歇的, 断续的) stream of visitors.
○	Poor: Long periods of inactivity with only a few periods of moderate activity.
Traffic Quality	
○	Good: Many booth visitors were decision-makers or decision-influencers.
○	Fair: About an equal mix of clerks and administrators.
○	Poor: Mostly clerks.
Visitor Reception	
○	Good: Most visitors appeared highly interested; they asked lots of questions and were clearly engaged with our product. Lots of enthusiastic comments.
○	Fair: Some visitors were interested and enthusiastic, but an equal number appeared only mildly interested.
○	Poor: Most visitors appeared uninterested; they didn't stay long at the booth, didn't ask many questions and quickly lost interest in our product.
Overall Show Evaluation	
○	Must-go next year.
○	Give it one more try.
○	Avoid next year.

Project

Project Guidelines

This project aims to go through the whole process of products distribution and after-sales service. The whole task is divided into three steps. Step One is about deciding on your company's marketing strategy in a specific area. Step Two focuses on the procedure of business negotiation between the salesmen and customers. Step Three centers on handling feedbacks from customers.

Please follow the *Task Description* to complete the project.

Task Description

Step One

- Organize a small group of 4-6 students in your class;
- Discuss with your group members about:
  a. the specific area where your products is to be launched;
  b. the strengths of your company's products;
  c. customers' needs in that specific area;
- Decide on your company's marketing strategy in that area.

Step Two

- Divide your group into two sides: one side being the salesmen, the other being the potential customers;
- Salesmen introduce the products to the customers, and the customers ask about quality and function of the products as well as the services to be offered;
- The two sides negotiate the prices, mode of payment and delivery of the products and reach an agreement.

Step Three

- Role-play the following situation: The customers make a complaint after they used the products for a period of time (the complaint may be about the quality of products or the services the company offers); and the salesmen deal with customers' complaint by offering solutions;
- The two sides express satisfaction with the solutions and show goodwill of maintaining long-term relationship with each other.

Self-evaluation

Rate your progress in this unit.	D	M	P	F*
I can work out marketing strategies for a company effectively.	☐	☐	☐	☐
I can use the marketing skills to introduce products to customers.	☐	☐	☐	☐
I can negotiate details of business with customers professionally.	☐	☐	☐	☐
I can deal with customers' complaints about the products and services.	☐	☐	☐	☐
I can collect customers' feedback on products and services.	☐	☐	☐	☐
I can write a reply to a letter of invitation to the fair.	☐	☐	☐	☐

***Note**: Distinction, Merit, Pass, Fail*

New Words and Expressions

Reading A

New Words

aggressive /əˈgrɛsɪv/ a. 进取的；挑衅的
cab /kæb/ n. 驾驶室
cultivate /ˈkʌltəˌvet/ v. 培养
durability /ˌdʊrəˈbɪləti/ n. 耐久性，耐用性
enhancement /ɪnˈhænsmənt/ n. 增进，增强
excavator /ˈɛkskəˌvetɚ/ n. 挖掘机
guarantee /ˌgærənˈti/ n. 保证，担保
in-house /ˌɪnˈhaʊs/ a. 内部的，室内的
niche /nɪtʃ/ n. 适当的位置，适合的处所
oriented /ˈɔrɪɛntɪd/ a. 导向的
prompt /prɑmpt/ a. 迅速的，立刻的
proof /pruf/ n. 证明，证据
reinforce /ˌriɪnˈfɔrs/ v. 加强，增强
release /rɪˈlis/ n. 发行，发布
stage /stedʒ/ v. 筹划，上演
topographical /ˌtɑpəˈgræfɪkl/ a. 地形学的
transparency /trænsˈpærənsi/ n. 透明度，透明性
visibility /ˌvɪzəˈbɪləti/ n. 能见度，可见性
workability /ˌwɝkəˈbɪləti/ n. 实用性，工作能力

Phrases & Expressions

appeal to 对（某人）有吸引力，使（某人）感兴趣
be central to 对……至关重要
distinguish... from... 把……和……区别开
industry association 行业协会
make efforts to 为……作出努力
make inroads into 进军，侵占
respond to 对……作出回应
specialize in 专门研究……
take advantage of 利用……优势

Technical Terms

wheel loader 轮式装载机

Proper Names

Doosan Infracore 斗山英维高株式会社

Reading B

New Words

awaited /əˈwetɪd/ a. 被期待的
convey /kənˈve/ v. 传达，传递
event /ɪˈvɛnt/ n. 大事，活动
hesitate /ˈhɛzəˌtet/ v. 犹豫
highlight /ˈhaɪˌlaɪt/ v. 使突出，使显著
hydraulics /haɪˈdrɔlɪks/ n. 液压机械
medium /ˈmidiəm/ n. 媒介，中介
proactive /proˈæktɪv/ a. 积极的
spatial /ˈspeʃəl/ a. 空间的
sponsor /ˈspɑnsɚ/ v. 赞助
venture /ˈvɛntʃɚ/ n. （商业）投机

Phrases & Expressions

on display 展出，展示
other than 除了
to one's advantage 对某人有利或有帮助

Proper Names

India Machine Tools Show 印度机械工具展
India Trade Promotion Organization 印度贸易促进组织
New Delhi 新德里

Vocabulary and Structure

Task 1 Write out the words in Reading A or Reading B according to their meanings in the right column. The first letters are already given.

h_____	to deal with, manage or control (people, a situation, a machine, etc.)
s_____	to arrange for something to take place
c_____	to acquire or develop a relationship
p_____	acting without delay
d_____	to show something to an interested audience
i_____	to fill somebody with thoughts, feelings or aims
r_____	to make stronger
h_____	to pause before saying or doing something because you are not sure

Task 2 Fill in each blank with the appropriate form of the word given in the brackets.

1. Future (expand) _____ of the business calls for two new factories.

2. In these hard times, many small firms have gone to the wall because they could not (competitive) _____ with big businesses.

3. No matter how modern and (comfort) _____ a hotel building is, the most important feature is the people who work there.

4. Governed by supply and demand, prices are the (visibility) _____ hand in economics.

5. Gas is one of the modern (convenient) _____ the newly-built apartment building provides.

6. Mr. Broackes died just two weeks after the (announce) _____ that I had won the 1993 Nobel Prize in Medicine.

7. My job is a dead end; I just work in the stores and there's absolutely no prospect of (promote) _____.

8. After the training, each (participate) _____ will take with them a computer and a set of VBI cards.

9. According to international norm, Chinese students who are (sponsor) _____ by the government to study abroad have the duty to return to serve their home country.

10. Without a moment's (hesitate) _____, she jumped into the river to save the drowning (溺水的) girl.

Task 3 Complete the following sentences with the words or phrases given below. Change the form if necessary.

respond to	surpass	remain	release	appeal to
on display	launch	highlight	convey	specialize in

1. His latest record was a big hit within a week of its _____.
2. We _____ offering clients a variety of Chinese merchandise at a competitive price.
3. Any car that would _____ these young customers had to have three main features: great styling, strong performance, and a low price.
4. During these quarterly meetings, it's the boss' responsibility to _____ each manager's plan.
5. When he put the shoes _____, they lasted no more than one hour.
6. In order to catch up with and _____ the advanced world levels we'll have to accelerate our speed.
7. Air at higher levels may _____ warmer than that near the ground.
8. If we _____ our new model a few weeks before the motor show we'll have the edge over our competitors.
9. Today her books, though no longer easy to get, still read with a freshness that only a good writer can _____.
10. This kind of dark material will _____ the brightness of your complexion.

Task 4 Make sentences with the same pattern as is shown in the examples.

> *Example 1:* the diamond ring shining brilliantly in the sun/his never-ending love for his wife
> — The diamond ring shining brilliantly in the sun is *a proof of* his never-ending love for his wife.
>
> *Example 2:* the fans' frantic screaming at the sight of the football star/his popularity among younger people
> — The fans' frantic screaming at the sight of the football star is *a proof* of his popularity among younger people.

1. smiles on the teacher's face/his satisfaction with the students' performance

2. his fingerprints on the criminal scene/his involvement in the murder that shocked the whole nation

3. the leaking faucet running all the time/the poor management of the hotel

4. gray hair and rough skin/the hardships she has experienced in recent years

5. the diploma in colored silk frame hung on the wall/her graduation from the best university in China

> *Example 1:* us/such an event has been running successfully in India over the past decade
> — *It is a matter of pride* for us that such an event has been running successfully in India over the past decade.
>
> *Example 2:* the whole nation/they had the quickest and best boat to cross the Atlantic in the Victorian Age (维多利亚时代)
> — *It was a matter of pride* for the whole nation that they had the quickest and best boat to cross the Atlantic in the Victorian Age.

1. all the Chinese basketball fans/Yao Ming took his team the Huston Rockets to the playoffs (<常规赛季之后的>季后赛) this year
   _____

2. Johnson/he depends on himself instead of turning to an adult when he needs help
   _____

3. the parents/their son was one of the three astronauts in the manned spacecraft Shenzhou-7
   _____

4. professor Mitch/Cambridge University recommended a candidate to his program
   _____

5. Liu Huan/he was invited to sing in the opening ceremony of the 2008 Olympic Games
   _____

Task 5　Translate the following sentences into English using the words or phrases given in the brackets.

1. Please come to my office next Monday _____ (到时你可以听到一些对你有利的消息). (to your advantage)

2. The manager emphasized _____ (进一步加强企业管理的重要性) in his annual report. (reinforce)

3. _____ (特区政府欢迎真正的企业家来港开展事业), bringing with them capital and expertise. (establish a presence)

4. _____ (改革的主要目标是帮助学生全面发展) so that they can meet the challenges of the new millennium. (primary objective)

5. As your agent, we will _____ (尽全力推销你方产品). (make effort to)

Grammar

Noun Clauses

Task 1 Underline the clauses in the following sentences and identify them as object, subject, predicative or appositive.

1. I believe that our team will win the basketball match.   (object)
2. Are you sure his answer is right?   (   )
3. We found it impossible that he could finish it in such a short time.   (   )
4. Everything depends on whether you agree with us.   (   )
5. That she becomes an artist may have been due to her father's influence.   (   )
6. What he says is not important.   (   )
7. My suggestion is that we should start early tomorrow.   (   )
8. At that time, it seemed as if I couldn't find the right word.   (   )
9. The news that Tom will go abroad is told by himself.   (   )
10. With the letter was his promise that he would visit me this Christmas.   (   )

Task 2 Circle the correct words to complete the statements or questions.

1. Please tell me where (does he live/(he lives)).
2. He is wondering (when can he/when he can) finish this difficult job.
3. He wanted to know (why she was/why was she) crying in the corner.
4. Would you tell me how much (you paid/did you pay) for the car?
5. Will you show how (this work is done/is this work done)?
6. Can you imagine what kind of man (is he/he is)?
7. Whether (he will/will he) accept the job is difficult to say.
8. That was what (did she say/she said) in the court yesterday.
9. The question is whether (the enemy is/is the enemy) marching toward us.
10. Has Mr. White made the decision that (should we/we should) send more people there?

Task 3 Tick the sentence that contains a noun clause.

1. (a) The news that he told me yesterday was true. ( )
   (b) The news that our team has won the game is true. (√)
2. (a) I had no idea that you were here. ( )
   (b) The idea that she had was crazy. ( )
3. (a) That question whether we need it has not been considered. ( )
   (b) The question that our teacher gave to us was difficult to answer. ( )
4. (a) The fact that we talked about is very important. ( )
   (b) The fact that the Chinese people invented the compass is known to all. ( )
5. (a) I never forget the day when I saw the Great Wall. ( )
   (b) The question when we will start the work is not decided. ( )
6. (a) The reply that he did not need help came as a surprise. ( )
   (b) The reply that contained important information was received too late. ( )

Task 4 Find out the noun clauses in the following paragraph and underline them.

   <u>Why some very good students often fail exams</u> was recently studied by a professor of psychology at New York University. Professor Iris Fodor conducted research on the anxiety of some students before taking exams. She stated that many students fail exams because they become extremely nervous and cannot think logically. Furthermore, although they have studied hard, they are afraid of whatever is on the exam. Extremely nervous students forget everything they have studied, and some even become sick before a test. According to Fodor, how a student feels before a test is very important. She worked with 50 students and taught them how they could reduce their test anxiety and perform well on their exams. What encouraged her was that most students in the program felt better able to cope with their anxiety. Her report *Test Anxiety Can Be Significantly Reduced* will be published in the University's newspaper and help more students to cope with exam anxiety.

名词性从句

名词性从句的功能相当于名词，它在复合句中能担任主语、宾语、表语、同位语、介词宾语等，因此根据它在句中不同的语法功能，名词性从句可分为主语从句、宾语从句、表语从句和同位语从句。

例 句	从 句
<u>What Jack did</u> shocked his friends. It is certain <u>that we will be late</u>.	主语从句
Billy's friends didn't know <u>that he couldn't swim</u>. I feel it a pity <u>that she can't come</u>. Mary is not responsible for <u>what Billy did</u>.	宾语从句
Tom's mistake was <u>that he refused to take lessons</u>.	表语从句
I have no idea <u>where he has gone</u>.	同位语从句

一、名词性从句引导词

主语从句：that; whether; who(m), what, which; whose; when, where, why, how等。
宾语从句：that; if, whether; who(m), what, which; whose; when, where, why, how等。
表语从句：that; whether; who(m), what, which; whose; when, where, why, how等。
同位语从句：that; whether; when, where, why等。

二、if和whether的区别

if和whether表示"是否"时，可以引导宾语从句，常常可以互换，但它们也有些区别。以下几种情况常用whether来引导从句，不用if。

1. 引导主语从句并在句首时，如：Whether we go there is not decided.
2. 引导表语从句时，如：The question is whether we can get there on time.
3. 引导同位语从句时，如：He asked me the question whether the work was worth doing.
4. 在介词后引导宾语从句时，如：I'm thinking about whether we'll have a meeting.
5. 直接与not连用时，如：I don't know whether or not you will go.
6. 宾语从句置于句首时，如：Whether you have met George before, I can't remember.

三、名词性从句的语序

含名词性从句的复合句无论是疑问句还是陈述句，无论名词性从句用什么引导词，其语序一律都用陈述语序。如：

Do you know what time it is?
Everybody wondered where Billy went.

四、that引导的同位语从句与定语从句的区别

引导同位语从句的that是连词，只起语法作用，在从句中不充当任何成分，但不能省略；而引导定语从句的that是关系代词，它除了起引导从句的语法作用之外，还要在从句中充当句子成分，主要作主语或宾语，作宾语时可省略。如：

The news that she had passed the exam made her parents very happy. (同位语从句)
The news (that) he told us interested all of us. (定语从句, that 可省略)

Comprehensive Exercises

Task 1 Fill in the blanks in the following sentences with proper subordinators.

1. Show us _how_ you do it.
2. Mary didn't say _____ she would be back.
3. You have to tell me _____ you didn't attend the lecture.
4. Football players don't care _____ it rains _____ not.
5. We must find out _____ did all this.
6. _____ you will fail is certain.
7. _____ caused the accident is still a mystery.
8. The trouble is _____ we don't have enough money.
9. The news _____ we are having a holiday tomorrow is not true.
10. There is some doubt _____ he will come.

Task 2 Change the underlined parts into noun clauses.

1. No one knows <u>the time of his coming</u>.
   No one knows <u>when he will come</u>.
2. I heard <u>of his success</u>.
   I heard _____.
3. We will never know <u>the reason for his failure</u>.
   We will never know _____.
4. Grandma insisted <u>on our staying with her</u>.
   Grandma insisted _____.
5. <u>The trouble</u> is that I have lost his address.
   _____ is that I have lost his address.
6. It is a great honor <u>for me to have been invited to the party</u>.
   It is a great honor _____.
7. The problem was <u>his failure to fulfill our expectations</u>.
   The problem was _____.
8. His only requirement is <u>the early completion of the system</u>.
   His only requirement is _____.
9. The report <u>of his retirement</u> was false.
   The report _____ was false.
10. News <u>of admission of more students into universities</u> has been put forward.
    News _____ has been put forward.

Task 3 Combine the pair of simple sentences into a complex sentence containing a noun clause.

1. He is an honest boy. Everybody knows it.
   <u>Everybody knows that he is an honest boy.</u>

2. He had made a mistake. He knew this.
   _____

3. A good environment is essential for a happy life. He believes this.
   _____

4. Could he depend on the guide? The traveler did not know.
   _____

5. The Earth moves around the sun. It is known to everybody.
   _____

6. Why did she buy the old car? It is a great mystery to us all.
   _____

7. Our class will go to a picnic this weekend. We have discussed the idea.
   _____

8. Picasso was a great artist. Nobody can challenge this fact.
   _____

9. The game was lost. It was the consequence of his carelessness.
   _____

10. We should start making preparations right now. My suggestion is that.
    _____

Task 4 Correct the errors that have been underlined in the following sentences.

1. My friend knows where do I live.
   My friend knows where I live.
2. I don't know what is your company address.
3. We never doubt whether he is honest.
4. Whomever was responsible for the accident is not yet clear.
5. If she comes or not doesn't concern me.
6. I feel it a terrible thing which my mother should undertake such a challenging task.
7. The fact is what he didn't notice the car until too late.
8. Your brother's health is not which it used to be.
9. The reason why he was late was whether he didn't catch the early bus.
10. There was little probability which they would succeed, but they didn't mind.

Fun Time

Why a Coca-Cola Salesman Failed in the Middle East

A disappointed salesman of Coca-Cola returns from his Middle East assignment.

A friend asked, "Why weren't you successful with the Arabs?"

The salesman explained, "When I got posted in the Middle East, I was very confident that I will make a good sales pitch (销售点) as Coca-Cola is virtually unknown there. But I had a problem that I didn't know how to speak Arabic. So I planned to convey the message through three posters...

First poster, a man is crawling through the hot desert sand... totally exhausted (筋疲力尽的) and panting (气喘吁吁的). Second poster, the man is drinking our Cola, and third, our man is now totally refreshed. Then these posters were posted all over the place."

"That should have worked," said the friend.

The salesman replied, "Well, not only did I not speak Arabic, but I didn't realize that Arabs read from right to left..."

Glossary

Words

A

		Unit
accelerate v.	促进, 加快……的速度	2A
accompany v.	伴随, 和……一起发生	3A
accumulate v.	积累, 积聚	7A
accurate a.	准确的, 精确的	2B
acid n.	（化）酸	6B
adaptation n.	适应, 改编	7A
additional a.	另外的, 附加的	7A
adjustable a.	可调整的, 可调校的	3A
aggressive a.	进取的；挑衅的	8A
aircraft n.	飞机, 飞行器	4A
alignment n.	校直, 调准	5B
amplify v.	放大, 增强	6A
annual a.	一年的, 全年的	2A
anti-corrosion n.	防侵蚀	6B
application n.	应用	2B
appropriate a.	适当的, 恰当的	1B
assemble v.	装配, 组合	3A
assistance n.	帮助, 援助	4A
associate v.	使发生联系, 联合	3A
attribute n.	属性, 特性	5A
automate v.	（使）自动化	7A
awaited a.	被期待的	8B

B

barrel n.	圆筒, 筒状物	4B
batch n.	一批, 成批（工作件）	5A
bearing n.	轴承	3B
board n.	板材	6A

C

		Unit
cab n.	驾驶室	8A
catalog n.	目录	1B
cavity n.	腔, 凹处	4B
centrifugal a.	离心的	2A
challenge n.	挑战性的要求, 难题	2B
chemical n.	化学制品	
a.	化学的	7B
chuck n.	卡盘	5B
civil a.	民用的	1B
classification n.	类别, 种类	1A
coded a.	编码的	4A
compactness n.	密实度, 紧密	3B
competitively ad.	有竞争力地	1A
complacent a.	自满的, 自鸣得意的	1A
component n.	组成部分, 元件	3A
compressor n.	压缩机, 压缩器	2A
compromise n./v.	妥协, 让步	1A
constantly ad.	不断地, 经常地	1A
construction n.	建筑（物）	1A
container n.	容器	4B
convert v.	转变, 转化	3B
convey v.	传达, 传递	8B
convince v.	使确信, 使信服	1B
cooperate v.	合作	1A
coordinate v.	协调	4A
cost-effective a.	有成本效益的, 划算的	7A
cultivate v.	培养	8A
currency n.	流通	7A

195

D

		Unit
database n.	数据库	4A
decoration n.	装饰, 装饰品	1A
defective n.	残次品	
a.	有缺陷的	5A
delivery n.	运送物	2A
density n.	密度, 稠密	3B
discipline n.	纪律	7B
dispatch n.	分派, 派遣	7B
disruption n.	中断, 破坏	7B
distinguish v.	区别, 使显出不同特征	7A
distribute v.	分布	2A
distribution n.	分配, 分发	3B
document v.	用文件记录	5B
draft n.	草图, 设计图	3A
drill v.	钻孔	4A
drug n.	药, 药物	7B
durability n.	耐久性, 耐用性	8A

E

edge n.	边, 棱, 边缘	3A
edit v.	编辑	3A
eliminate v.	消除	6A
enable v.	使能够, 使可能	3A
enclose v.	装入, 圈起, 封闭	3B
enhancement n.	增进, 增强	8A
ensure v.	确保, 保证	1A
essential a.	基本的	1B
essentially ad.	本质上, 根本地	4A
estimate v.	评估, 测量	5A
event n.	大事, 活动	8B
excavator n.	挖掘机	8A
exceed v.	超过	2A
exist v.	存在(某种情况)	2A
expand v.	扩大, 扩展	1A
export v.	出口	1A

		Unit
extensive a.	大范围的, 大量的	2B
external a.	外部的, 外面的	1A
extract v.	摘出, 选取	4A

F

facilitate v.	推动, 促进	7A
fierce a.	激烈的, 极度的	2A
file n.	锉刀	6B
finding n. (pl.)	结果	5B
formula n.	配方, 公式	1A
foundation n.	基础, 根本	7A
furnace n.	熔炉	4B

G

gear n.	齿轮, 传动装置	3B
gearbox n.	变速箱, 变速器	3B
generality n.	通性, 普遍(性)	4A
geometric a.	几何的, 几何学的	3A
graphical a.	绘成图表的, 图解形式的	5A
grease n.	润滑油	6B
groove n.	长而窄的槽	6B
guarantee n.	保证, 担保	8A

H

harden v.	变硬, 凝固	4B
handheld a.	手持式的, 便携的	7B
hardware n.	五金器具, (电子仪器的) 部件	6A
headstock n.	头架	5B
hesitate v.	犹豫	8B
highlight v.	使突出, 使显著	8B
hollow a.	中空的	4B
horizontal a.	水平的	5B

Glossary

		Unit
housing *n.*	外壳, 外罩	3B
hub *n.*	轮轴, 轮毂	4B
hydraulics *n.*	液压机械	8B
hypothesis *n.*	假设	5A

I

implement *v.*	完成, 实现	3A
indicator *n.*	指标	2A
infer *v.*	推断, 推论	5A
in-house *a.*	内部的, 室内的	8A
innovation *n.*	革新, 创新	7A
innovative *a.*	创新的, 革新的	2B
insert *v.*	插入, 嵌入	3A
inspection *n.*	检查	5A
inspector *n.*	检查员	5B
integrated *a.*	集成的, 综合的	1A
intend *v.*	计划, 打算	5A
interface *n.*	接口, 接合处	4A
interpret *v.*	解释, 翻译	4A
inventory *n.*	库存	4A
item *n.*	项目, 条款	1B

J

jar *n.*	瓶, 罐	4B

L

limitation *n.*	局限, 限制	2B
load *v.*	加载, 装入程序	4A
localization *n.*	地方化, 本土化	2A
long-standing *a.*	久经考验的, 经久不衰的	1B
low-end *a.*	低端的, 低档的	3A

M

		Unit
machine *v.*	以机器制造	4A
manually *ad.*	手动	3A
manufacturer *n.*	生产商, 制造商	1A
mass *n.*	大量	4B
maximize *v.*	最佳化, 最大化	7A
medium *n.*	媒介, 中介	8B
metallurgy *n.*	冶金	2A
mill *n.*	压榨机, 磨粉机	6A
minimize *v.*	减少到最低限度	6A
model *n.*	型号, 样式	1B
modification *n.*	更改, 改变, 修改	3A
molten *n.*	熔化的, 熔融状的	4B
mount *v.*	安装; 安置	6A
multi-lube *n.*	多功能润滑油/润滑剂	6B
multiply *v.*	增加, 乘	3B

N

nail *n.*	指甲, 钉子	6B
niche *n.*	适当的位置, 适合的处所	8A
numerical *a.*	数字的	4A
nut *n.*	螺母, 螺帽	6A

O

optimize *v.*	使最优化	7A
orbit *v.*	环绕轨道运行	3B
oriented *a.*	导向的	8A
original *a.*	最初的, 最早的	1A
output *n.*	产量	2A
outstanding *a.*	出色的	3B
outward *ad.*	向外	4B
overcome *v.*	战胜, 克服(某事物)	2B

P

		Unit
parameter n.	参数, 参量	5A
parametric a.	参数的, 参量的	3A
party n.	一方, 当事人	1B
paste n.	糊状物, 膏	6B
penetrate v.	穿透	6B
perimeter n.	周边, 周长	4A
perishable a.	容易腐烂的	7B
petrochemical n.	石油化学制品	2A
pinion n.	小齿轮	3B
piston n.	活塞	2A
placement n.	布局	3A
planet n.	行星	3B
platform n.	平台	6A
polish v.	磨光, 擦亮	6B
population n.	(统计学中) 母体, 总体	5A
powder n.	粉末	4B
preferable a.	更可取的, 更好的	7A
pre-finished a.	抛光好的	6A
previous a.	(时间或顺序上) 先的, 前的	2A
previously ad.	以前, 先前	4A
primary a.	首要的, 主要的	5B
principal a.	主要的, 最重要的	5A
proactive a.	积极的	8B
process n.	程序, 流程	1A
professional a.	专业的, 职业的	1A
prompt a.	迅速的, 立刻的	8A
proof n.	证明, 证据	8A
property n.	属性, 特性	7A
purchase n./v.	购买	1B

Q

quote v.	报价	7A

R

		Unit
rail n.	轨道	6B
randomly ad.	随机地	5A
reciprocate v.	(指机件) 沿直线往复移动	2A
recommend v.	推荐	1B
recruitment n.	招募, 招聘	7B
reference n.	证明文书	1A
regulation n.	规章, 规则	7B
reinforce v.	加强, 增强	8A
relatively ad.	相对地	2A
release n.	发行, 发布	8A
reliability n.	可靠 (reliable是其形容词形式)	1B
remark n.	评论, 备注	5B
representative n.	代表	1B
residue n.	残余, 滤渣	6B
resin n.	树脂	6B
resolve v.	解决 (问题、疑问等)	2B
reveal v.	显露出, 展现出	4B
revenue n.	收益	2A
review v.	审查, 回顾	4A
rotational a.	旋转的	4B
RPM (abbr.) revolutions per minute	转数/分	3B
runout n.	偏斜, 偏心率	5B
rusty a.	生锈的, 腐蚀的	6B

S

sample n.	范例, 样品	1B
scale n.	规模, 范围	3A
scenario n.	预料或期望的一系列事件模式	7A
screw n.	螺丝钉	6A

Glossary

		Unit
secure *v.*	确保	6A
semi-conducted *a.*	半引导的	2B
semiconductor *n.*	半导体	2B
sensor *n.*	传感器	5A
sequence *n.*	系,一连串	4A
shaft *n.*	轴	3B
signal *n.*	信号,指令	5A
sincere *a.*	真诚的	1B
sketch *n.*	草图,素描	
v.	描绘略图	3A
slate *n.*	石板,石片	7A
solar *a.*	太阳的	3B
solid *n.*	实体	3A
solution *n.*	溶液	6B
spatial *a.*	空间的	8B
specialist *n.*	专家	2B
spindle *n.*	主轴	6B
spoke *n.*	轮辐	4B
sponsor *v.*	赞助	8B
spot *n.*	现场	1A
spray *v.*	喷洒,喷射	4B
spray *n.*	喷雾	6B
staff *n.*	(全体)职员,工作人员	7B
stage *v.*	筹划,上演	8A
statistical *a.*	统计学的	5A
stiffness *n.*	坚硬,硬度	3B
stock *n.*	库存	7B
storage *n.*	贮藏	6A
subsidiary *a.*	附属的,副的	1A
subtract *v.*	减去,扣除	3A
supplier *n.*	供应商,厂商	1B
swing *v.*	使旋转,摆动	4B

T

		Unit
tailstock *n.*	尾架,尾座	5B
textile *n.*	纺织	2A
thermoplastic *n.*	热塑性塑料	1A
tip *v.*	变倾斜	6A
tolerance *n.*	(偏离标准的)容许误差,公差	4A
topographical *a.*	地形学的	8A
torque *n.*	(机器的)扭转力,扭矩,转矩	3B
transmit *v.*	传动,传输	3B
transparency *n.*	透明度,透明性	8A
tube *n.*	管,管状物	4B

U

		Unit
unique *a.*	独一无二的,独特的	1A
update *v.*	更新	7A
utensil *n.*	器皿,用具	4B

V

		Unit
validity *n.*	有效性,合法性	7A
variability *n.*	变量,变率	5A
vary *v.*	变化,不同	7B
venture *n.*	(商业)投机	8B
version *n.*	版本	7A
vertical *a.*	垂直的	5B
vertically *ad.*	垂直地	4B
vibration *n.*	振动	6A
virtual *a.*	虚拟的	3A
visibility *n.*	能见度,可见性	8A
vitally *ad.*	极为,生死攸关地	1B

W

		Unit
washer *n.*	垫圈	6A
wedge *n.*	楔	6B
what-if *n.*	假定分析	7A
wireframe *n.*	线框，框架	3A
workability *n.*	实用性，工作能力	8A

		Unit
workbench *n.*	工作台，如机械工人、木匠或珠宝匠的工作台	6A
workflow *n.*	工作流程	7A

Phrases & Expressions

A

		Unit
act as	担任, 充当	6A
a high proportion of	很大一部分	3B
allow for	考虑到, 为了……	6A
amount to	总计	2A
appeal to	对 (某人) 有吸引力, 使 (某人) 感兴趣	8A

B

be about to	将要, 正打算	1B
be bound to	一定要, 一定会	6B
be central to	对……至关重要	8A
be committed to	致力于	1A
be compared with	与……相比较	3A
be defined as	被定义为	4A
be engaged in	专营于	1B
be in control	在控制之中	5A
be likely to	很有可能, 倾向于	7B
be responsible for	对……负责任	2B
be secured to	固定在, 安装在	6A

C

center around	以……为中心, 围绕	4A
check... in	核对无误	7B
conform to	和……一致	4B
cooperate with	与……合作, 与……共同努力	1A
custom parts	客户定制/特需的部件	7A

D

		Unit
distinguish... from...	把……和……区别开	8A
due to	由于, 因为	3B

F

fit in	相合	6B

G

go out of control	不受控制	5A

H

head office	总部	1B

I

in accordance with	与……一致	5B
industry association	行业协会	8A
in other words	换句话说	3B
in principle	原则上, 基本上	3A
in stock	库存, 现货	7B
integrated PDM system	集成PDM系统	7A
in terms of	根据, 按照, 用……的话, 在……方面	1B
in the field of	在……领域	1A
in this respect	在这方面	1B
invest... on	投资于	1B

201

K

		Unit
keep track of	记录, 跟踪	7B

L

lay out	设计, 布局	3A

M

make contributions to	对……做出贡献	2B
make efforts to	为……作出努力	8A
make inroads into	进军, 侵占	8A

O

on display	展出, 展示	8B
on no account	决不	6B
other than	除了	8B

P

pan out	结果好, 成功	7A
pick out	挑出	5A
prior to	在前, 居先	1B

Q

		Unit
put away	放好	6A

R

refer to... as	把……称作, 把……当作	3A
rely on	依赖, 依靠	4A
respond to	对……作出回应	8A
roll back	回转	7A

S

sink in	渗入, 浸透	6B
specialize in	专门研究……	8A

T

take advantage of	利用……优势	8A
to one's advantage	对某人有利或有帮助	8B

W

with washers and nuts	带着垫圈和螺帽	6A

Technical Terms

		Unit
3D dumb solid	三维块体	3A
3D parametric solid modeling	三维参数化实体建模	3A

A

acceptance sampling	进料抽样试验	5A

C

centrifugal gas compressor	离心式压缩机	2A
contact inspection methods	接触检测方法	5A
control chart	管理图表	5A

D

design specifications	设计规格	5B

E

extrusion mold	挤压式模具	1A

F

fine inching control	精密缓动控制	1A

G

gas compressor	气体压缩机	2A
guide rail	导轨	6B

L

		Unit
lifting gear	吊运装置	7B

M

mass properties	质量属性	3A
measured value	测量值	5B
mechanical probe	机械探针	5A
MEMS Micro-Electro-Mechanical systems	微机电系统	2B
MEMS switch	微机电系统开关	2B

N

noncontact inspection methods	非接触检测方法	5A

O

oil residue	油渣	6B

P

physical quantity	物理量	5A
piston gas compressor	活塞式压缩机	2A
precision bench lathe	精密台式车床	5B
PVC low foam technology	PVC (Polyvinyl Chloride聚氯乙烯) 低发泡技术	1A

R

reciprocating gas compressor	往复式压缩机	2A

S

solid volume	实体积	3A
spindle center	主轴中心	5B

		Unit
spindle nose	主轴端	5B
spindle thread	螺纹轴	6B
statistical quality control	统计质量管理	5A

T

test of hypothesis	假设检测	5A
two-dimensional projected views	二维投影视图	3A

W

wheel loader	轮式装载机	8A
WPC raw material formula	WPC (木塑复合) 原材料配方	1A

Proper Names

		Unit
Doosan Infracore	斗山英维高株式会社	8A
India Machine Tools Show	印度机械工具展	8B
India Trade Promotion Organization	印度贸易促进组织	8B
National Australia Bank (NAB)	澳大利亚国立银行	1B
New Delhi	新德里	8B
Sydney	悉尼	1B